Fig. 1.

The Solar System.

W Carl Rufus.

METEORIC ASTRONOMY:

A TREATISE

ON

SHOOTING-STARS, FIRE-BALLS,

AND

AEROLITES.

BY

DANIEL KIRKWOOD, LL.D.
PROFESSOR OF MATHEMATICS IN WASHINGTON AND JEFFERSON COLLEGE.

PHILADELPHIA:
J. B. LIPPINCOTT & CO.
[illegible]

PREFACE.

Aristotle and other ancient writers regarded comets as meteors generated in the atmosphere. This opinion was generally accepted, even by the learned, until the observations of Tycho, near the close of the sixteenth century, showed those mysterious objects to be more distant than the moon, thus raising them to the dignity of *celestial* bodies. An achievement somewhat similar, and certainly no less interesting, was reserved for the astronomers of the *nineteenth* century. This was the great discovery that *shooting-stars, fire-balls, and meteoric stones, are, like comets, cosmical bodies moving in conic sections about the sun.* Dr. Halley was the first to foretell the return of a comet, and the year 1759 will ever be known in history as that which witnessed the fulfillment of his prophecy. But in the department of *meteoric* astronomy, a similar honor must now be awarded to the late Dr. Olbers. Soon after the great star-shower of 1833 he inferred from a comparison of recorded facts that the November display attains a maximum at intervals of thirty three or thirty-four years. He accordingly designated 1866 or 1867 as the time of its probable return; and the night of November 13th of the former year must always be memorable as affording the first verification of *his* prediction. On that night several thousand meteors were observed in one hour from a single station. This remarkable display, together with the fact that another still more brilliant is looked for in November, 1867, has given meteoric astronomy a more than ordinary degree of interest in the public mind. To gratify, in some measure, the curiosity which has been awakened, by presenting in a popular form the principal results of observation and study in this new field of research, is the main design of the following work.

The first two chapters contain a popular view of what is known in regard to the star-showers of August and November, and also of some other epochs. The third is a description, in chronological order, of the most important falls of meteoric stones, together with the phenomena attending their descent. The fourth and following chapters to the eleventh inclusive, discuss various questions in the theory of meteors: such, for instance, as the relative number of aerolitic falls during different parts of the day, and also of the year; the coexistence of the different forms of meteoric matter in the same rings; meteoric dust; the stability of the solar system; the doctrine of a resisting medium; the extent of the atmosphere as indicated by meteors; the meteoric theory of solar heat; and the phenomena of variable and temporary stars. The twelfth chapter regards the rings of Saturn as dense meteoric swarms, and accounts for the principal interval between them. The thirteenth presents various facts, not previously noticed, respecting the asteroid zone between Mars and Jupiter, with suggestions concerning their cause or explanation

As the nebular hypothesis furnishes a plausible account of the origin of meteoric streams, it seemed desirable to present an intelligible view of that celebrated theory. This accordingly forms the subject of the closing chapter.

The greater part of the following treatise, it is proper to remark, was written before the publication (in England) of Dr. Phipson's volume on "Meteors, Aerolites, and Falling-stars." The author has had that work before him, however, while completing his manuscript, and has availed himself of some of the accounts there given of recent phenomena.

CANONSBURG, PA., *May*, 1867.

CONTENTS.

CHAPTER VIII.

CHAPTER IX.

CHAPTER X.

CHAPTER XI.

CHAPTER XII.

CHAPTER XIII.

CHAPTER XIV.

INTRODUCTION.

A GENERAL VIEW OF THE SOLAR SYSTEM.

THE SOLAR SYSTEM consists of the sun, together with the planets and comets which revolve around him as the center of their motions. The sun is the great controlling orb of this system, and the source of light and heat to its various members. Its magnitude is one million four hundred thousand times greater than that of the earth, and it contains more than seven hundred times as much matter as all the planets put together.

MERCURY is the nearest planet to the sun ; its mean distance being about thirty-seven millions of miles. Its diameter is about three thousand miles, and it completes its orbital revolution in 88 days.

VENUS, the next member of the system, is sometimes our morning and sometimes our evening star. Its magnitude is almost exactly the same as that of the earth. It revolves round the sun in 225 days.

THE EARTH is the third planet from the sun in the order of distance; the radius of its orbit being about ninety-five millions of miles. It is attended by one satellite—the moon—the diameter of which is 2160 miles.

MARS is the first planet exterior to the earth's orbit. It is considerably smaller than the earth, and has no satellite. It revolves round the sun in 687 days.

THE ASTEROIDS.—Since the commencement of the present century a remarkable zone of telescopic planets has been discovered immediately exterior to the orbit of Mars. These bodies are extremely small; some of them probably containing less matter than the largest mountains on the earth's surface. More than ninety members of the group are known at present, and the number is annually increasing.

JUPITER, the first planet exterior to the asteroids, is nearly five hundred millions of miles from the sun, and revolves round him in a little less than twelve years. This planet is ninety thousand miles in diameter and contains more than twice as much matter as all the other planets, primary and secondary, put together. Jupiter is attended by four moons or satellites.

SATURN is the seventh planet in the order of distance—counting the asteroids as one. Its orbit is about four hundred millions of miles beyond that of Jupiter. This planet is attended by eight satellites, and is surrounded by three broad, flat rings. Saturn is seventy-six thousand miles in diameter, and its mass or quantity of matter is more than twice that of all the other planets except Jupiter.

URANUS is at double the distance of Saturn, or nineteen times that of the earth. Its diameter is about thirty-five thousand miles, and its period of revolution, eighty-four years. It is attended by four satellites.

NEPTUNE is the most remote known member of the system; its distance being nearly three thousand millions of miles. It is somewhat larger than Uranus; has certainly one satellite, and probably several more. Its period is about one hundred and sixty-five years. A cannon-ball flying at the rate of five hundred miles per hour would not reach the orbit of Neptune from the sun in less than six hundred and eighty years.

These planets all move round the sun in the same direction—from west to east. Their motions are nearly circular, and also nearly in the same plane. Their orbits, except that of Neptune, are represented in the frontispiece. It is proper to remark, however, that all representations of the solar system by maps and planetariums must give an exceedingly erroneous view either of the magnitudes or distances of its various members. If the earth, for instance, be denoted by a ball half an inch in diameter, the diameter of the sun, according to the same scale (sixteen thousand miles to the inch), will be between four and five feet; that of the earth's orbit, about one thousand feet; while that of Neptune's orbit will be nearly six miles. To give an accurate representation of the solar system at a single view is therefore plainly impracticable.

COMETS.—The number of comets belonging to our system is unknown. The appearance of more than seven hundred has been recorded, and of this number, the elements of about two hundred have been computed. They move in very eccentric orbits—some, perhaps, in parabolas or hyperbolas.

THE ZODIACAL LIGHT is a term first applied by Dominic Cassini, in 1683, to a faint nebulous aurora, somewhat resembling the milky-way, apparently of a conical or lenticular form, having its base toward the sun, and its axis nearly in the direction of the ecliptic. The most favorable time for observing it is when its axis is most nearly perpendicular to the horizon. This, in our latitudes, occurs in March for the evening, and in October for the morning. The angular distance of its vertex from the sun is frequently seventy or eighty degrees, while sometimes, though rarely (except within the tropics), it exceeds even one hundred degrees.

The zodiacal light is probably identical with the meteor called *trabes* by *Pliny* and *Seneca*. It was noticed in the latter part of the sixteenth century by Tycho Brahé, who "considered it to be an abnormal spring-evening twilight."

It was described by Descartes about the year 1630, and again by Childrey in 1661. The first accurate description of the phenomenon was given, however, by Cassini. This astronomer supposed the appearance to be produced by the blended light of an innumerable multitude of extremely small planetary bodies revolving in a ring about the sun. The appearance of the phenomenon as seen in this country is represented in Fig. 2.

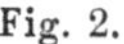

Fig. 2.

For general readers it may not be improper to premise the following explanations:

Meteors are of two kinds, *cosmical* and *terrestrial:* the former traverse the interplanetary spaces; the latter originate in the earth's atmosphere.

Bolides is a general name for meteoric fire-balls of greater magnitude than shooting-stars.

The *period* of a planet, comet, or meteor is the time which it occupies in completing one orbital revolution.

The motion of a heavenly body is said to be *direct* when it is from west to east; and *retrograde* when it is from east to west.

Encke's Hypothesis of a Resisting Medium.—The time occupied by Encke's comet in completing its revolution about the sun is becoming less and less at each successive return. Professor Encke explains this fact by supposing the interplanetary spaces to be filled with an extremely rare fluid, the resistance of which to the cometary motion produces the observed contraction of the orbit.

METEORIC ASTRONOMY.

CHAPTER I.

SHOOTING-STARS.

I. The Meteors of November 12th–14th.

ALTHOUGH shooting-stars have doubtless been observed in all ages of the world, they have never, until recently, attracted the special attention of scientific men. The first exact observations of the phenomena were undertaken, about the close of the last century, by Messrs. Brandes and Benzenberg. The importance, however, of this new department of research was not generally recognized till after the brilliant meteoric display of November 13th, 1833. This shower of fire can never be forgotten by those who witnessed it.* The display was observed from the West Indies to British America, and from 60° to 100° west longitude from Greenwich. Captain Hammond, of the ship Restitution, had just arrived at Salem, Massachusetts, where he observed the phenomenon from midnight till daylight. He

* For a full description, see Silliman's Journal for January and April, 1834 (Prof. Olmsted's article). Also a valuable paper, in the July No. of the same year, by Prof. Twining.

noticed with astonishment that precisely one year before, viz., on the 13th of November, 1832, he had observed a similar appearance (although the meteors were less numerous) at Mocha, in Arabia. It was soon found, moreover, as a further and most remarkable coincidence, that an extraordinary fall of meteors had been witnessed on the 12th of November, 1799. This was seen and described by Andrew Ellicott, Esq., who was then at sea near Cape Florida. It was also observed in Cumana, South America, by Humboldt, who states that it was "simultaneously seen in the new continent, from the equator to New Herrnhut, in Greenland (lat. 64° 14′), and between 46° and 82° longitude."

This wonderful correspondence of dates excited a very lively interest throughout the scientific world. It was inferred that a recurrence of the phenomenon might be expected, and accordingly arrangements were made for systematic observations on the 12th, 13th, and 14th of November. The periodicity of the shower was thus, in a very short time, placed wholly beyond question. The examination of old historical records led to the discovery of at least 12 appearances of the November shower previous to the great fall of 1833. The descriptions of these phenomena will be found collected in an interesting article by Prof. H. A. Newton, in the *American Journal of Science and Arts*, for May, 1864. They occurred in the years 902, 931, 934, 1002, 1101, 1202, 1366, 1533, 1602, 1698, 1799, and 1832. Besides these 12 enumerated by Professor Newton as "the predecessors of the great exhibition on the morning of November 13th, 1833," we find 6 others, less distinctly marked, in the catalogue of M. Quetelet.* These were in the years 1787, 1818, 1822, 1823, 1828, and 1831. From 1833 to 1849, inclusive, Quetelet's catalogue indicates 11 partial returns of the November shower; making in all, up to the latter date, 29. In 1835, November 13th, a straw roof was set on fire by a meteoric fire-ball, in the department de l'Aine, France. On the 12th of November, 1837, "at 8 o'clock in the evening, the attention of observers in various parts of Great Britain was directed to a bright luminous body, apparently proceeding from the North, which, after making a rapid descent, in the manner of a rocket, suddenly burst, and scattering its particles into various beautiful forms, vanished in the atmosphere. This was succeeded by others all similar to the first, both in shape and the manner of its ultimate disappearance. The whole display terminated at ten o'clock, when dark clouds, which continued up till a late hour, overspread the earth, preventing any further observations."—*Milner's Gallery of Nature*, p. 142.

In 1838, November 12th–13th, meteors were observed in unusual numbers at Vienna. One of extraordinary brilliancy, having an apparent magnitude equal to that of the full moon, was seen near Cherburg.

On several other returns of the November epoch the number of meteors observed has been greater than on ordinary nights; the distinctly marked exhibitions, however, up to 1866, have all been enumerated.

---

* Physique du Globe, Chap. IV.

The Shower of November 14, 1866.

The fact that all great displays of the November meteors have taken place at intervals of thirty-three or thirty-four years, or some multiple of that period, had led to a general expectation of a brilliant shower in 1866. In this country, however, the public curiosity was much disappointed. The numbers seen were greater than on ordinary nights, but not such as would have attracted any special attention. The greatest number recorded at any one station was seen at New Haven, by Prof. Newton. On the night of the 12th, 694 were counted in five hours and twenty minutes, and on the following night, 881 in five hours. This was about six times the ordinary number. A more brilliant display was, however, witnessed in Europe. Meteors began to appear in unusual frequency about eleven o'clock on the night of the 13th, and continued to increase with great rapidity for more than two hours; the maximum being reached a little after one o'clock. The Edinburgh *Scotsman*, of November 14th, contains a highly interesting description of the phenomenon as observed at that city. "Standing on the Calton Hill, and looking westward," the editor remarks,—"with the Observatory shutting out the lights of Prince's Street—it was easy for the eye to delude the imagination into fancying some distant enemy bombarding Edinburgh Castle from long range; and the occasional cessation of the shower for a few seconds, only to break out again with more numerous and more brilliant drops of fire, served to countenance this fancy. Again, turning eastward, it was possible now and then to catch broken glimpses of the train of one of the meteors through the grim dark pillars of that ruin of most successful manufacture, the National Monument; and in fact from no point in or out of the city was it possible to watch the strange rain of stars, pervading as it did all points of the heavens, without pleased interest, and a kindling of the imagination, and often a touch of deeper feeling that bordered on awe. The spectacle, of which the loftiest and most elaborate description could but be at the best imperfect—which truly should have been seen to be imagined—will not soon pass from the memories of those to whose minds were last night presented the mysterious activities and boundless fecundities of that universe of the heavens, the very unchangeableness of whose beauty has to many made it monotonous and of no interest."

The appearance of the phenomenon, as witnessed at London, is minutely described in the *Times* of November 15th. The shower occurred chiefly between the hours of twelve and two. About one o'clock a single observer counted 200 in two minutes. The whole number seen at Greenwich was 8485. The shower was also observed in different countries on the continent.

The Meteors of 1866 *compared with those of former Displays.*

The star shower of 1866 was much inferior to those of 1799 and 1833.* With these exceptions, however,

* Professor Olmsted estimated the number of meteors, visible at New Haven, during the night of November 12th–13th, 1833, at 240,000.

it has, perhaps, been scarcely surpassed during the last 500 years. Historians represent the meteors of 902 as innumerable, and as moving like rain in all possible directions.* The exhibition of 1202 was no less magnificent. The stars, it is said, were seen to dash against each other like swarms of locusts; the phenomenon lasting till daybreak.† The shower of 1366 is thus described in a Portuguese chronicle, quoted by Humboldt: "In the year 1366, twenty-two days of the month of October being past, three months before the death of the king, Dom Pedro (of Portugal), there was in the heavens a movement of stars, such as men never before saw or heard of. At midnight, and for some time after, all the stars moved from the east to the west; and after being collected together, they began to move, some in one direction, and others in another. And afterward they fell from the sky in such numbers, and so thickly together, that as they descended low in the air, they seemed large and fiery, and the sky and the air seemed to be in flames, and even the earth appeared as if ready to take fire. That portion of the sky where there were no stars, seemed to be divided into many parts, and this lasted for a long time."

The following is Humboldt's description of the

---

* Conde says, "there were seen, as it were lances, an infinite number of stars, which scattered themselves like rain to the right and left, and that year was called 'the year of stars.'"

† In 1202, "on the last day of Muharrem, stars shot hither and thither in the heavens, eastward and westward, and flew against one another like a scattering swarm of locusts, to the right and left; this phenomenon lasted until daybreak; people were thrown into consternation, and cried to God the Most High with confused clamor." —Quoted by Prof. Newton, in Silliman's Journal, May, 1864.

shower of 1799, as witnessed by himself and Bonpland, in Cumana, South America: "From half after two, the most extraordinary luminous meteors were seen toward the east. . . . . Thousands of bolides and falling stars succeeded each other during four hours. They filled a space in the sky extending from the true east 30° toward the north and south. In an amplitude of 60° the meteors were seen to rise above the horizon at E. N. E. and at E., describe arcs more or less extended, and fall toward the south, after having followed the direction of the meridian. Some of them attained a height of 40°, and all exceeded 25° or 30°. . . . . Mr. Bonpland relates, that from the beginning of the phenomenon there was not a space in the firmament equal in extent to three diameters of the moon, that was not filled at every instant with bolides and falling-stars. . . . . The Guaiqueries in the Indian suburb came out and asserted that the firework had begun at one o'clock. . . . . The phenomenon ceased by degrees after four o'clock, and the bolides and falling-stars became less frequent; but we still distinguished some toward the northeast a quarter of an hour after sunrise."

Discussion of the Phenomena.

Since the memorable display of November 13th, 1833, the phenomena of shooting-stars have been observed and discussed by Brandes, Benzenberg, Olbers, Saigey, Heis, Olmsted, Herrick, Twining, Newton, Greg, and many others. In the elaborate paper of Professor Olmsted, it was shown that the meteors had their origin at a distance of more than

2000 miles from the earth's surface; that their paths diverged from a common point near the star *Gamma Leonis;* that in a number of instances they became visible about 80 miles from the earth's surface; that their velocity was comparable to that of the earth in its orbit; and that in some cases their extinction occurred at an elevation of 30 miles. It was inferred, moreover, that they consisted of combustible matter which took fire and was consumed in passing through the atmosphere; that this matter was derived from a nebulous body revolving round the sun in an elliptical orbit, but little inclined to the plane of the ecliptic; that its aphelion was near that point of the earth's orbit through which we annually pass about the 13th of November—the perihelion being a little within the orbit of Mercury; and finally that its period was about one-half that of the earth. Dr. Olmsted subsequently modified his theory, having been led by further observations to regard the zodiacal light as the nebulous body from which the shooting-stars are derived. The latter hypothesis was also adopted by the celebrated Biot.

The fact that the position of the radiant point does not change with the earth's rotation, places the cosmical origin of the meteors wholly beyond question. The theory of a closed ring of nebulous matter revolving round the sun in an elliptical orbit which intersects that of the earth, affords a simple and satisfactory explanation of the phenomena. This theory was adopted by Humboldt, Arago, and others, shortly after the occurrence of the meteoric shower of 1833. That the body which furnishes the material of these meteors moves in a closed or elliptical orbit is evident from the periodicity of the shower. It is also manifest from the partial recurrence of the phenomenon from year to year, that the matter is diffused around the orbit; while the extraordinary falls of 1833, 1799, 1366, and 1202, prove the diffusion to be far from uniform.

Elements of the Orbit.

Future observations, it may be hoped, will ultimately lead to an accurate determination of the elements of this ring: many years, however, will probably elapse before all the circumstances of its motion can be satisfactorily known. Professor Newton, of Yale College, has led the way in an able discussion of the observations.* He has shown that the different parts of the ring are, in all probability, of very unequal density; that the motion is retrograde; and that the time, during which the meteors complete a revolution about the sun, must be limited to one of five accurately determined periods, viz.: 180·05 days, 185·54 days, 354·62 days, 376·5 days, or 33·25 years. He makes the inclination of the ring to the ecliptic about 17°. The five periods specified, he remarks, "are not all equally probable. Some of the members of the group which visited us last November [1863] gave us the means of locating approximately the central point of the region from which the paths diverge. Mr. G. A. Nolen has, by graphical processes specially devised for the purpose, found its longitude to be 142°, and its latitude 8° 30′. This longitude

---

* Am. Journ. of Sci. and Arts, May and July, 1864.

is very nearly that of the point in the ecliptic toward which the earth is moving. Hence the point from which the absolute motion of the bodies is directed (being in a great circle through the other two points) has the same longitude. The absolute motion of each meteor, then, is directed very nearly at right angles to a line from it to the sun, the deviation being probably not more than two or three degrees.

"Now, if in one year the group make $2\pm\frac{1}{33\cdot25}$ revolutions, there is only a small portion of the orbit near the aphelion which fulfills the above condition. In like manner, if the periodic time is 33·25 years, only a small portion of the orbit near the perihelion fulfills it. On the other hand, if the annual motion is $1\pm\frac{1}{33\cdot25}$ revolutions, the required condition is answered through a large part of the orbit. Inasmuch as no reason appears why the earth should meet a group near its apsides rather than elsewhere, we must regard it as more probable that the group makes in one year either $1+\frac{1}{33\cdot25}$, or $1-\frac{1}{33\cdot25}$ revolutions."

Professor Newton concludes that the third of the above-mentioned periods, viz., 354·62 days, combines the greatest amount of probability of being the true one. We grant the force of the reasons assigned for its adoption. At least one consideration, however, in favor of the long period of 33·25 years is by no means destitute of weight: of nearly 100 known bodies which revolve about the sun in orbits of small eccentricity, not one has a retrograde motion. Now if this striking fact has resulted from a general cause, how shall we account for the backward motion of a meteoric ring, in an orbit almost circular, and but little inclined to the plane of the ecliptic? In such a case, is not the preponderance of probability in favor of the longer period?

A revolution in 33·25 years corresponds to an ellipse whose major axis is 20·6. Consequently the aphelion distance would be somewhat greater than the mean distance of Uranus. It may also be worthy of note, that five periods of the ring would be very nearly equal to two of Uranus.

The *Monthly Notices of the Royal Astronomical Society* for December, 1866, and January, 1867, contain numerous articles on the star shower of November 13th–14th, 1866. Sir John Herschel carefully observed the phenomena, and his conclusions in regard to the orbit are confirmatory of those of Professor Newton. "We are constrained to conclude," he remarks, "that the true line of direction, in space of each meteor's flight, lay in a plane at right angles to the earth's radius vector at the moment; and that therefore, except in the improbable assumption that the meteor was at that moment *in perihelio* or *in aphelio*, its orbit would not deviate greatly from the circular form." The question is one to be decided by observation, and the only meteor whose track and time of flight seem to have been well observed, is that described by Professor Newton in *Silliman's Journal* for January, 1867, p. 86. The velocity in this case, if the estimated time of flight was nearly correct, was *inconsistent with the theory of a circular orbit.*

It is also worthy of notice that Dr. Oppolzer's elements of the first comet of 1866 resemble, in a remarkable manner, those of the meteoric ring, supposing the latter to have a period of about 33¼ years.

Schiaparelli's elements of the November ring, and Oppolzer's elements of the comet of 1866, are as follows:

	November Meteors.	Comet of 1866.
Longitude of perihelion...............	56° 25′	60° 28′
Longitude of ascending node..........	231 28	231 26
Inclination................................	17 44	17 18
Perihelion distance......................	0·9873	0·9765
Eccentricity	0·9046	0·9054
Semi-axis major..........................	10·3400	10·3240
Period, in years..........................	33·2500	33·1760
Motion..	Retrograde.	Retrograde.

It seems very improbable that these coincidences should be accidental. Leverrier and other astronomers have found elements of the meteoric orbit agreeing closely with those given by Schiaparelli. Should the identity of the orbits be fully confirmed, it will follow that the comet of 1866 *is a very large meteor* of the November stream.

The researches of Professor C. Bruhns, of Leipzig, in regard to this group of meteors afford a probable explanation of the division of Biela's comet—a phenomenon which has greatly perplexed astronomers for the last twenty years. Adopting the period of 33¼ years, Professor Bruhns finds that the comet passed extremely near, and probably *through* the meteoric ring near the last of December, 1845. It is easy to perceive that such a collision might produce the separation soon afterward observed.

As the comet of Biela makes three revolutions in twenty years, it was again at this intersection, or approximate intersection of orbits about the end of 1865. But although the comet's position, with re-

Fig. 3.

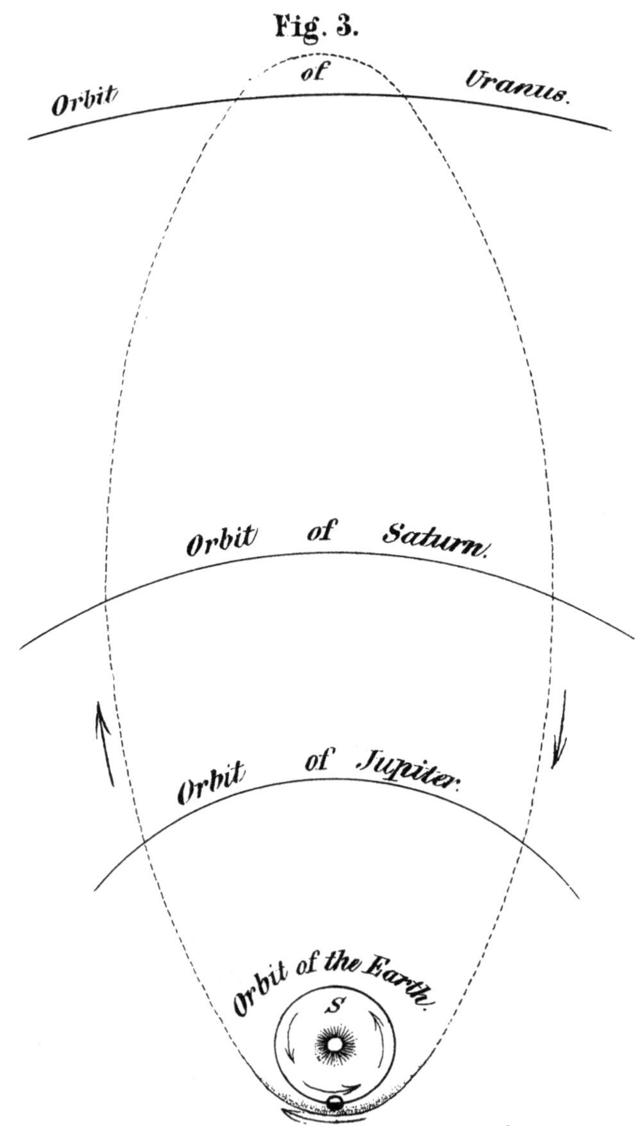

Probable Orbit of the November Meteors.

spect to the earth, was the same as in 1845–6, and although astronomers watched eagerly for its appearance, their search was unsuccessful. In short, *the comet is lost.* The denser portion of the meteoric stream was then approaching its perihelion. A portion of the arc had even passed that point, as a meteoric shower was observed at Greenwich on the 13th of November, 1865.* The motion of the meteoric stream is retrograde; that of the comet, direct. Did the latter plunge into the former, and was its non appearance the result of such collision and entanglement?

---

* The stream or arc of meteors is several years in passing its node. The first indication of the approach of the display of 1866 was the appearance of meteors in unusual numbers at Malta, on the 13th of November, 1864. The great length of the arc is indicated, moreover, by the showers of 931 and 934.

CHAPTER II.

OTHER METEORIC RINGS.

II. The Meteors of August 6th–11th.

MUSCHENBROEK, in his *Introduction to Natural Philosophy*, published in 1762, called attention to the fact that shooting-stars are more abundant in August than in any other part of the year. The annual periodicity of the maximum on the 9th or 10th of the month was first shown, however, by Quetelet, shortly after the discovery of the yearly return of the November phenomenon. Since that time an extraordinary number of meteors has been regularly observed, both in Europe and America, from the 7th to the 11th of the month; the greatest number being generally seen on the 10th. In 1839, Edward Heis, of Aix-la-Chapelle, saw 160 meteors in one hour on the night of the 10th. In 1842, he saw 34 in ten minutes at the time of the maximum. In 1861, on the night of the 10th, four observers, watching together at New Haven, saw in three hours—from ten to one o'clock—289 meteors. On the same night, at Natick, Massachusetts, two observers saw 397 in about seven hours. At London, Mercer County, Pennsylvania, on the night of August 9th, 1866, Samuel S. Gilson, Esq., watching alone, saw 72 meteors in forty minutes, and, with an assistant,

117 in one hour and fifteen minutes. Generally, the number observed per hour, at the time of the August maximum, is about nine times as great as on ordinary nights. Like the November meteors, they have a common "radiant;" that is, their tracks, when produced backward, meet, or nearly meet, in a particular point in the constellation Perseus.

Of the 315 meteoric displays given in Quetelet's "Catalogue des principales apparitions d'étoiles filantes," 63 seem to have been derived from the August ring. The first 11 of these, with one exception, were observed in China during the last days of July, as follows:

1	A.D. 811,	July 25th.
2	820,	" 25th–30th.
3	824,	" 26th–28th.
4	830,	" 26th.
5	833,	" 27th.
6	835,	" 26th.
7	841,	" 25th–30th.
8	924,	" 27th–30th.
9	925,	" 27th–30th.
10	926,	" 27th–30th.
11	933,	" 25th–30th.

The next dates are 1243, August 2d, and 1451, August 7th. A comparison of these dates indicates a forward motion of the node of the ring along the ecliptic. This was pointed out several years since by Boguslawski. A similar motion of the node has also been found in the case of the November ring. That these points should be stationary is, indeed, altogether improbable. The nodes of all the planetary orbits, it is well known, have a secular variation.

On the evening of August 10th, 1861, at about

11h. 30m., a meteor was seen by Mr. E. C. Herrick and Prof. A. C. Twining, at New Haven, Connecticut, which "was much more splendid than Venus, and left a train of sparks which remained luminous for twenty seconds after the meteor disappeared." The same meteor was also accurately observed at Burlington, New Jersey, by Mr. Benjamin V. Marsh. It was "conformable,"—that is, its track produced backward passed through the common radiant—and it was undoubtedly a member of the August group. The observations were discussed by Professor H. A. Newton, of Yale College, who deduced from them the following approximate elements of the ring:*

Semi-axis major	0 84
Eccentricity	0·28
Perihelion distance	0·60
Inclination	84°
Period	281 days.
Motion, retrograde.	

The earth moving at the rate of 68,000 miles per hour, is at least five days in passing entirely through the ring. This gives a thickness of more than 8,000,000 miles.

The result of Professor Newton's researches on the oribit of this ring, though undertaken with inadequate data, and hence, in some respects, probably far from correct, is nevertheless highly interesting as being the first attempt to determine the orbit of shooting-stars. More recent investigations have shown a remarkable resemblance between the elements of these meteors and those of the third comet

* Silliman's Journ. for Sept. and Nov., 1861.

of 1862. The former, by Schiaparelli, and the latter, by Oppolzer, are as follows:

	Meteors of August 10th.	Comet III., 1862.
Longitude of perihelion	343° 38′	344° 41
Ascending node..........	138 16	137 27
Inclination.................	63 3	66 25
Perihelion distance......	0·9643	0·9626
Period.......................	105 years (?).	123 years (?).
Motion	Retrograde.	Retrograde.

This similarity is too great to be accidental. *The August meteors and the third comet of* 1862 *probably belong to the same ring.*

III. The Meteors of April 18th–26th.

The following dates of the April meteoric showers are extracted from Quetelet's table previously referred to:

1..........	A.D. 401,	April	9th.	8...	A.D. 1096,	April	10th.
2..........	538,	"	7th.	9...	1122,	"	11th.
3..........	839,	"	17th.	10...	1123,	"	11th.
4..........	927,	"	17th.	11...	1803,	"	20th.
5..........	934,	"	18th.	12...	1838,	"	20th.
6..........	1009,	"	16th.	13...	1841,	"	19th.
7..........	1094,	"	10th.	14...	1850,	"	11th–17th.

The display of 401 was witnessed in China, and is described as "very remarkable." That of 1803 was best observed in Virginia, and was at its maximum between one and three o'clock. The alarm of fire had called many of the inhabitants of Richmond from their houses, so that the phenomenon was generally witnessed. The meteors "seemed to fall from every point in the heavens, in such numbers as to resemble a shower of sky-rockets." Some were of extraordinary magnitude. "One in particular, appeared to fall from the zenith, of the apparent size of a ball 18 inches in diameter, that lighted the whole hemisphere for several seconds."

The probability that the meteoric falls about the 20th of April are derived from a ring which intersects the earth's orbit, was first suggested by Arago, in 1836. The preceding list indicates a forward motion of the node. The radiant, according to Mr. Greg, is about *Corona.* The number of meteors observed in 1838, 1841, and 1850, was not very extraordinary. Recent observations indicate April 9th–12th as another epoch. The radiant is in Virgo.

IV. The Meteors of December 6th–13th.

On the 13th of December, 1795, a large meteoric stone fell in England. On the night, between the 6th and 7th of December, 1798, Professor Brandes, then a student in Göttingen, saw 2000 shooting-stars. On the 11th of the month, 1836, a fall of meteoric stones, described by Humboldt as "enormous," occurred near the village of Macao, in Brazil. During the last few years unusual numbers of shooting-stars have been noticed by different observers from the 10th to the 13th; the maximum occurring about the 11th. From A.D. 848, December 2d, to 1847, December 8th–10th, we find 14 star showers in Quetelet's catalogue, derived, probably, from this meteoric stream. As in other cases, the dates seem to show a progressive motion of the node. The position of the radiant, as determined by Benjamin V. Marsh,

Esq., of Philadelphia, from observations in 1861 and 1862, and also by R. P. Greg, Esq., of Manchester, England, is at a point midway between Castor and Pollux.

V. The Meteors of January 2d–3d.

About the middle of the present century, Mr. Julius Schmidt, of Bonn, a distinguished and accurate observer, designated the 2d of January as a meteoric epoch; characterizing it, however, as "probably somewhat doubtful." Recent observations, especially those of R. P. Greg, Esq., have fully confirmed it. The meteors for several hours are said to be as numerous as at the August maximum. The radiant is near the star *Beta* of the constellation Böotes.

Quetelet's list contains at least five exhibitions which belong to this epoch. Two or three others may also be referred to it with more or less probability.

Several other meteoric epochs have been indicated; some of which, however, must yet be regarded as doubtful. In thirty years, from 1809 to 1839, 12 falls of bolides and meteoric stones occurred from the 27th to the 29th of November. Such coincidences can hardly be accidental. Unusual numbers of shooting-stars have also been seen about the 27th of July; from the 15th to the 19th of October, and about the middle of February. The radiant, for the last-mentioned epoch, is in *Leo Minor*. The numbers observed in October are said to be at present increasing. At least seven of the exhibitions in Quetelet's catalogue are referable to this epoch. It is worthy of remark, moreover, that three of the dates specified by Mr. Greg as *aerolite* epochs are coincident with those of shooting-stars; viz., February 15th–19th, July 26th, and December 13th. The whole number of exhibitions enumerated in Quetelet's catalogue is 315. In eighty-two instances the day of the month on which the phenomenon occurred is not specified. Nearly two-thirds of the remainder, as we have seen, belong to established epochs, and the periodicity of others will perhaps yet be discovered. But reasons are not wanting for believing that our system is traversed by numerous meteoric streams besides those which actually intersect the earth's orbit. The asteroid region between Mars and Jupiter is probably occupied by such an annulus. The number of these asteroids increases as their magnitudes diminish; and this doubtless continues to be the case far below the limit of telescopic discovery. The zodiacal light is probably a dense meteoric ring, or rather, perhaps, a number of rings. We speak of it as *dense* in comparison with others, which are invisible except by the ignition of their particles in passing through the atmosphere. From a discussion of the motions of the perihelia of Mercury and Mars, Leverrier has inferred the existence of two rings of minute asteroids; one within the orbit of Mercury, whose mass is nearly equal to that of Mercury himself; the other at the mean distance of the earth, whose mass cannot *exceed* the tenth part of the mass of the earth.

Within the last few years a distinguished European savant, Buys-Ballot, of Utrecht, has discovered a short period of variation in the amount of solar heat received by the earth: the time from one maximum to another exceeding the period of the sun's apparent rotation by about twelve hours. The variation cannot therefore be due to any inequality in the heating power of the different portions of the sun's surface. The discoverer has suggested that it may be produced by a meteoric ring, whose period slightly exceeds that of the sun's rotation. Such a zone might influence our temperature by partially intercepting the solar heat.

General Remarks.

1. The average number of shooting-stars seen in a clear, moonless night by a single observer, is about 8 per hour. *One* observer, however, sees only about one-fourth of those visible from his point of observation. About 30 per hour might therefore be seen by watching the entire hemisphere. In other words, 720 shooting-stars per day could be seen by the naked eye at any one point of the earth's surface, did the sun, moon, and clouds permit.

2. The mean altitude of shooting-stars above the earth's surface is about 60 miles.

3. The number visible over the whole earth is about 10,460 times the number to be seen at any one point. Hence the average number of those daily entering the atmosphere and having sufficient magnitude to be seen by the naked eye, is about 7,532,600.

4. The observations of Pape and Winnecke indicate that the number of meteors visible through the telescope, employed by the latter, is about 53 times

the number visible to the naked eye, or about 400,000,000 per day.* This is two per day, or 73,000 per century, for every square mile of the earth's surface. By increasing the optical power, this number would probably be indefinitely increased. At special times, moreover, such as the epochs of the great meteoric showers, the addition of foreign matter to our atmosphere is much greater than ordinary. It becomes, therefore, an interesting question whether sensible changes may not thus be produced in the atmosphere of our planet.

5. In August, 1863, 20 shooting-stars were doubly observed in England; that is, they were seen at two different stations. The average weight of these meteors, estimated—in accordance with the mechanical theory of heat—from the quantity of light emitted, was a little more than two ounces.

6. A meteoric mass exterior to the atmosphere, and consequently non-luminous, was observed on the evening of October 4th, 1864, by Edward Heis, a distinguished European astronomer. It entered the field of view as he was observing the milky way, and he was enabled to follow it over 11 or 12 degrees of its path. It eclipsed, while in view, a number of the fixed stars.

---

* The numerical results here given are those found by Professor Newton. See Silliman's Journ. for March, 1865.

CHAPTER III.

AEROLITES.

It is now well known that much greater variety obtains in the structure of the solar system than was formerly supposed. This is true, not only in regard to the magnitudes and densities of the bodies composing it, but also in respect to the forms of their orbits. The whole number of planets, primary and secondary, known to the immortal author of the *Mecanique Celeste*, was only 29. This number has been more than quadrupled in the last quarter of a century. In Laplace's view, moreover, all comets were strangers within the solar domain, having entered it from without. It is now believed that a large proportion originated in the system and belong properly to it.

The gradation of planetary magnitudes, omitting such bodies as differ but little from those given, is presented at one view in the following table:

Name.	Diameter in miles.
Jupiter	90,000
Uranus	35,000
The Earth	7,926
Mercury	3,000
The Moon	2,160
Rhea, Saturn's 5th satellite	1,200
Dione " 4th "	500

Name.	Diameter in miles.
Vesta*	260
Juno	104
Melpomene	52
Polyhymnia	35
Isis	25
Atalanta	20
Hestia	15

The diminution doubtless continues indefinitely below the present limit of optical power. If, however, the orbits have small eccentricity, such asteroids could not become known to us unless their mean distances were nearly the same with that of the earth. But from the following table it will be seen that the variety is no less distinctly marked in the forms of the orbits:

Name.	Eccentricity.
Venus	0·00683
The Earth	0·01677
Jupiter	0·04824
Metis	0·12410
Mercury	0·20562
Pallas	0·24000
Polyhymnia	0·33820
Faye's comet	0·55660
D'Arrest's "	0·66090
Biela's "	0·75580
Encke's "	0·84670
Halley's "	0·96740
Fourth comet of 1857	0·98140
Fifth comet of 1858 (Donati's)	0·99620
Third comet of 1827	0·99927

Were the eccentricities of the nearest asteroids equal to that of Faye's comet, they would in perihelion

* The diameters of the asteroids are derived from a table by Prof. Lespiault, in the Rep. of the Smithsonian Inst. for 1861, p. 216.

intersect the earth's orbit. Now, in the case of both asteroids and comets, the smallest are the most numerous; and as this doubtless continues below the limit of telescopic discovery, the earth ought to encounter such bodies in its annual motion. *It actually does so.* The number of *cometoids* thus encountered in the form of *meteoric stones*, *fire-balls*, and *shooting-stars* in the course of a single year amounts to many millions. The extremely minute, and such as consist of matter in the gaseous form, are consumed or dissipated in the upper regions of the atmosphere. No deposit from ordinary shooting-stars has ever been known to reach the earth's surface. But there is probably great variety in the physical constitution of the bodies encountered; and though comparatively few contain a sufficient amount of matter in the solid form to reach the surface of our planet, scarcely a year passes without the fall of meteoric stones in some part of the earth, either singly or in clusters. Now, when we consider how small a proportion of the whole number are probably observed, it is obvious that the actual occurrence of the phenomenon can be by no means rare.*

Although numerous instances of the fall of aerolites had been recorded, some of them apparently well authenticated, the occurrence long appeared too marvelous and improbable to gain credence with scientific men. Such a shower of rocky fragments occurred, however, on the 26th of April, 1803, at

* "It appears probable, from the researches of Schreibers, that 700 fall annually."—Cosmos, vol. i. p. 119 (Bohn's Ed.). Reichenbach makes the number much greater.

L'Aigle, in France, as forever to dissipate all doubt on the subject. At one o'clock P.M., the heavens being almost cloudless, a tremendous noise, like that of thunder, was heard, and at the same time an immense fire-ball was seen moving with great rapidity through the atmosphere. This was followed by a violent explosion which lasted several minutes, and which was heard not only at L'Aigle, but in every direction around it to the distance of seventy miles. Immediately after a great number of meteoric stones fell to the earth, generally penetrating to some distance beneath the surface. The largest of these fragments weighed 17½ pounds. This occurrence very naturally excited great attention. M. Biot, under the authority of the government, repaired to L'Aigle, collected the various facts in regard to the phenomenon, took the depositions of witnesses, etc., and finally embraced the results of his investigations in an elaborate memoir.

It would not comport with the design of the present treatise to give an extended list of these phenomena. The following account, however, includes the most important instances of the fall of aerolites, and also of the displays of meteoric fire-balls.

1. According to Livy a number of meteoric stones fell on the Alban Hill, near Rome, about the year 654 B.C. This is the most ancient fall of aerolites on record.

2. 468 B.C., about the year in which Socrates was born. A mass of rock, described as "of the size of two millstones," fell at Ægos Potamos, in Thrace. An attempt to rediscover this meteoric mass, so celebrated in antiquity, was recently made, but with-

out success. Notwithstanding this failure, Humboldt expressed the hope that, as such a body would be difficult to destroy, it may yet be found, "since the region in which it fell is now become so easy of access to European travelers."

3. 921 A.D. An immense aerolite fell into the river (a branch of the Tiber) at Narni, in Italy. It projected three or four feet above the surface of the water.

4. 1492, November 7th. An aerolite, weighing two hundred and seventy-six pounds, fell at Ensisheim, in Alsace, penetrating the earth to the depth of three feet. This stone, or the greater portion of it, may still be seen at Ensisheim.

5. 1511, September 14th. At noon an almost total darkening of the heavens occurred at Crema. "During this midnight gloom," says a writer of that period, "unheard-of thunders, mingled with awful lightnings, resounded through the heavens. * * * On the plain of Crema, where never before was seen a stone the size of an egg, there fell pieces of rock of enormous dimensions and of immense weight. It is said that ten of these were found weighing a hundred pounds each." A monk was struck dead at Crema by one of these rocky fragments. This terrific meteoric display is said to have lasted two hours, and 1200 aerolites were subsequently found.

6. 1637, November 29th. A stone, weighing fifty-four pounds, fell on Mount Vaison, in Provence.

7. 1650, March 30th. A Franciscan monk was killed at Milan by the fall of a meteoric stone.

8. 1674. Two Swedish sailors were killed on shipboard by the fall of an aerolite.

9. 1686, July 19th. An extraordinary fire-ball was seen in England; its motion being opposite to that of the earth in its orbit. Halley pronounced this meteor a cosmical body. (See Philos. Transact., vol. xxix.)

10. 1706, June 7th. A stone weighing seventy-two pounds fell at Larissa, in Macedonia.

11. 1719, March 19th. Another great meteor was seen in England. Its explosion occurred at an elevation of 69 miles. Notwithstanding its height, however, the report was like that of a broadside, and so great was the concussion that windows and doors were violently shaken.

12. 1751, May 26th. Two meteoric masses, consisting almost wholly of iron, fell near Agram, the capital of Croatia. The larger fragment, which weighs seventy-two pounds, is now in Vienna.

13. 1756. The concussion produced by a meteoric explosion threw down chimneys at Aix, in Provence, and was mistaken for an earthquake.

14. 1771, July 17th. A large meteor exploded near Paris, at an elevation of 25 miles.

15. 1783, August 18th. A fire-ball of extraordinary magnitude was seen in Scotland, England, and France. It produced a rumbling sound like distant thunder, although its elevation above the earth's surface was 50 miles at the time of its explosion. The velocity of its motion was equal to that of the earth in its orbit, and its diameter, according to Sir Charles Blagden, was about half a mile.

16. 1790, July 24th. Between nine and ten o'clock at night a very large igneous meteor was seen near Bourdeaux, France. Over Barbotan a loud explosion

was heard, which was followed by a shower of meteoric stones of various magnitudes.

17. 1794, July. A fall of about a dozen aerolites occurred at Sienna, Tuscany.

18. 1795, December 13th. A large meteoric stone fell near Wold Cottage, in Yorkshire, England. The following account of the phenomenon is taken from Milner's *Gallery of Nature*, p. 134: "Several persons heard the report of an explosion in the air, followed by a hissing sound; and afterward felt a shock, as if a heavy body had fallen to the ground at a little distance from them. One of these, a plowman, saw a huge stone falling toward the earth, eight or nine yards from the place where he stood. It threw up the mould on every side; and after penetrating through the soil, lodged some inches deep in solid chalk rock. Upon being raised, the stone was found to weigh fifty-six pounds. It fell in the afternoon of a mild but hazy day, during which there was no thunder or lightning; and the noise of the explosion was heard through a considerable district."

19. 1796, February 19th. A stone of ten pounds' weight fell in Portugal.

20. 1798, March 12th. A stone weighing twenty pounds fell at Sules, near Ville Franche.

21. 1798, March 17th. An aerolite weighing about twenty pounds fell at Sale, Department of the Rhone.

22. 1798, December 19th. A shower of meteoric stones fell at Benares, in the East Indies. An interesting account of the phenomenon was given by J. Lloyd Williams, F.R.S., then a resident in Bengal. The sky had been perfectly clear for several days. At eight o'clock in the evening a large meteor ap-

peared, which was attended with a loud rumbling noise. Immediately after the explosion a sound was heard like that of heavy bodies falling in the neighborhood. Next morning the fresh earth was found turned up in many places, and aerolites of various sizes were discovered beneath the surface.

23. 1803, April 26th. The shower at L'Aigle, previously described.

24. 1807, December 14th. A large meteor exploded over Weston, Connecticut. The height, direction, velocity, and magnitude of this body were ably discussed by Dr. Bowditch in a memoir communicated to the American Academy of Arts and Sciences in 1815. The following condensed statement of the principal facts, embodied in Dr. Bowditch's paper, is extracted from the *People's Magazine* for January 25th, 1834:

"The meteor of 1807 was observed about a quarter-past six on Monday morning. The day had just dawned, and there was little light except from the moon, which was just setting. It seemed to be half the diameter of the full moon; and passed, like a globe of fire, across the northern margin of the sky. It passed behind some clouds, and when it came out it flashed like heat lightning. It had a train of light, and appeared like a burning fire-brand carried against the wind. It continued in sight about half a minute, and, in about an equal space after it faded, three loud and distinct reports, like those of a four-pounder near at hand, were heard. Then followed a quick succession of smaller reports, seeming like what soldiers call a running fire. The appearance of the meteor was as if it took three successive throes, or leaps, and at each explosion a rushing of stones was heard through the air, some of which struck the ground with a heavy fall.

"The first fall was in the town of Huntington, near the house of Mr. Merwin Burr. He was standing in the road, in front of his house, when the stone fell, and struck a rock of granite about fifty feet from him, with a loud noise. The rock was stained a dark-red color, and the stone was principally shivered into very small fragments, which were thrown around to a distance of twenty feet. The largest piece was about the size of a goose egg, and was still warm.

"The stones of the second explosion fell about five miles distant, near Mr. William Prince's residence, in Weston. He and his family were in bed when they heard the explosion, and also heard a heavy body fall to the earth. They afterward found a hole in the earth, about twenty-five feet from the house, like a newly dug post-hole, about one foot in diameter, and two feet deep, in which they found a meteoric stone buried, which weighed thirty-five pounds. Another mass fell half a mile distant, upon a rock, which it split in two, and was itself shivered to pieces. Another piece, weighing thirteen pounds, fell a half a mile to the northeast, into a plowed field.

"At the last explosion, a mass of stone fell in a field belonging to Mr. Elijah Seely, about thirty rods from the house. This stone falling on a ledge, was shivered to pieces. It plowed up a large portion of the ground, and scattered the earth and stones to the distance of fifty or a hundred feet. Some cattle that were near were very much frightened, and jumped into an inclosure. It was concluded that this last stone, before being broken, must have weighed about two hundred pounds. These stones were all of a similar nature, and different from any commonly found on this globe. When first found, they were easily reduced to powder by the fingers, but by exposure to the air they gradually hardened."

25. 1859, November 15th. Between nine and ten o'clock in the morning, an extraordinary meteor was seen in several of the New England States, New York, New Jersey, the District of Columbia, and Virginia. The apparent diameter of the head was nearly equal to that of the sun, and it had a train, notwithstanding the bright sunshine, several degrees in length. Its disappearance on the coast of the Atlantic was followed by a series of the most terrific explosions. It is believed to have descended into the water, probably into Delaware Bay. A highly interesting account of this meteor, by Prof. Loomis, may be found in the *American Journal of Science and Arts* for January, 1860.

26. 1860, May 1st. About twenty minutes before one o'clock P.M., a shower of meteoric stones—one of the most extraordinary on record—fell in the S. W. corner of Guernsey County, Ohio. Full accounts of the phenomena are given in *Silliman's Journal* for July, 1860, and January and July, 1861, by Professors E. B. Andrews, E. W. Evans, J. L. Smith, and D. W. Johnson. From these interesting papers we learn that the course of the meteor was about 40° west of north. Its visible track was over Washington and Noble Counties, and the prolongation of

its projection, on the earth's surface, passes directly through New Concord, in the S. E. corner of Muskingum County. The height of the meteor, when seen, was about 40 miles, and its path was nearly parallel with the earth's surface. The sky, at the time, was, for the most part, covered with clouds over northwestern Ohio, so that if any portion of the meteoric mass continued on its course, it was invisible. The velocity of the meteor, in relation to the earth's surface, was from 3 to 4 miles per second; and hence its absolute velocity in the solar system was from 20 to 21 miles per second. This would indicate an orbit of considerable eccentricity.

"At New Concord,* Muskingum County, where the meteoric stones fell, and in the immediate neighborhood, there were many distinct and loud reports heard. At New Concord there were first heard in the sky, a little southeast of the zenith, a loud detonation, which was compared to that of a cannon fired at the distance of half a mile. After an interval of ten seconds another similar report. After two or three seconds another, and so on with diminishing intervals. Twenty-three distinct detonations were heard, after which the sounds became blended together and were compared to the rattling fire of an awkward squad of soldiers, and by others to the roar of a railway train. These sounds, with their reverberations, are thought to have continued for two minutes. The last sounds seemed to come from a point in the southeast 45° below the zenith. The

* New Concord is close to the Guernsey County line. Nearly all the stones fell in Guernsey.

result of this cannonading was the falling of a large number of stony meteorites upon an area of about ten miles long by three wide. The sky was cloudy, but some of the stones were seen first as 'black specks,' then as 'black birds,' and finally falling to the ground. A few were picked up within twenty or thirty minutes. The warmest was no warmer than if it had lain on the ground exposed to the sun's rays. They penetrated the earth from two to three feet. The largest stone, which weighed one hundred and three pounds, struck the earth at the foot of a large oak tree, and, after cutting off two roots, one five inches in diameter, and grazing a third root, it descended two feet ten inches into hard clay. This stone was found resting under a root that was not cut off. This would seemingly imply that it entered the earth obliquely."

Over thirty of the stones which fell were discovered, while doubtless many, especially of the smaller, being deeply buried beneath the soil, entirely escaped observation. The weight of the largest ten was four hundred and eighteen pounds.

27. 1864, May 14th. Early in the evening a very large and brilliant meteor was seen in France, from Paris to the Spanish border. At Montauban, and in the vicinity, loud explosions were heard, and showers of meteoric stones fell near the villages of Orgueil and Nohic. The principal facts in regard to this meteor are the following:

Elevation when first seen, over..................	55 miles.
" at the time of its explosion.........	20 "
Inclination of its path to the horizon..........	20° or 25°
Velocity per second, about........................	20 miles,

or equal to that of the earth's orbital motion. "This example," says Prof. Newton, "affords the strongest proof that the detonating and stone-producing meteors are phenomena not essentially unlike."

The foregoing list contains but a small proportion even of those meteoric stones the date of whose fall is known. But besides these, other masses have been found so closely similar in structure to aerolites whose descent has been observed, as to leave no doubt in regard to their origin. One of these is a mass of iron and nickel, weighing sixteen hundred and eighty pounds, found by the traveler Pallas, in 1749, at Abakansk, in Siberia. This immense aerolite may be seen in the Imperial Museum at St. Petersburg. On the plain of Otumpa, in Buenos Ayres, is a meteoric mass 7½ feet in length, partly buried in the ground. Its estimated weight is thirty-three thousand six hundred pounds. A specimen of this stone, weighing fourteen hundred pounds, has been removed and deposited in one of the rooms of the British Museum. A similar block, of meteoric origin, weighing twelve or thirteen thousand pounds, was discovered some years since in the Province of Bahia, in Brazil.

Some of the inferences derived from the examination of meteoric stones, and the consideration of the phenomena attending their fall, are the following:

1. R. P. Greg, Esq., of Manchester, England, who has made luminous meteors a special study, has found that meteoric stone-falls occur with greater frequency than usual on or about particular days. He calls attention especially to five aerolite epochs, viz.:

February 15th–19th; May 19th; July 26th; November 29th, and December 13th.

2. It is worthy of remark that no new elements have been found in meteoric stones. Humboldt, in his *Cosmos*, called attention to this interesting fact. "I would ask," he remarks, "why the elementary substances that compose one group of cosmical bodies, or one planetary system, may not in a great measure be identical? Why should we not adopt this view, since we may conjecture that these planetary bodies, like all the larger or smaller agglomerated masses revolving round the sun, have been thrown off from the once far more expanded solar atmosphere, and have been formed from vaporous rings describing their orbits round the central body?"*

3. But while aerolites contain no elements but such as are found in the earth's crust, the manner in which these elements are combined and arranged is so peculiar that a skillful mineralogist will readily distinguish them from terrestrial substances.

4. Of the eighteen or nineteen elements hitherto observed in meteoric stones, iron is found in the greatest abundance. The specific gravities vary from 1·94 to 7·901: the former being that of the stone of Alais, the latter, that of the meteorite of Wayne County, Ohio, described by Professor J. L. Smith in *Silliman's Journal* for November, 1864, p. 385. In most cases, however, the specific gravity is about 3 or 4.

5. The contemplation of the heavenly bodies has

---

* Cosmos, vol. i. p. 120.

often produced in thoughtful minds an intense desire to know something of their nature and physical constitution. This curiosity is gratified in the examination of aerolites. To handle, weigh, inspect, and analyze bodies that have wandered unnumbered ages through the planetary spaces—perhaps approaching in their perihelia within a comparatively short distance of the solar surface, and again receding in their aphelia to the limits of the planetary system—must naturally excite a train of pleasurable emotions.

6. It is highly probable that in pre-historic times, before the solar system had reached its present stage of maturity, those chaotic wanderers were more numerous in the vicinity of the earth's orbit than in recent epochs. Even now the interior planets, Mercury and Venus, appear to be moving through the masses of matter which constitute the zodiacal light. It would seem probable, therefore, that they are receiving from this source much greater accretions of matter than the earth.

7. As Mercury's orbit is very eccentric, he is beyond his mean distance during much more than half his period. Hence, probably, the greater increments of meteoric matter are derived from such portions of the zodiacal light as have a longer period than Mercury himself. If so, the tendency would be to diminish slowly the planet's mean motion. Such a lengthening of the period has been actually discovered.*

* Leverrier's Annals of the Observatory of Paris, vol. i. p. 38.

CHAPTER IV.

CONJECTURES IN REGARD TO METEORIC EPOCHS.

It is highly probable that aerolites and shooting-stars are derived either from rings thrown off in the planes of the solar or planetary equators, or from streams of nebulous matter drawn into the solar system by the sun's attraction. Such annuli or streams would probably each furnish an immense number of meteor-asteroids. If any rings intersect the earth's orbit, our planet must encounter such masses as happen at the same time to be passing the point of intersection. This must be repeated *at the same epoch* in different years; the frequency of the encounter of course depending on the closeness and regularity with which the masses are distributed around the ring. Accordingly it has been found that not only the meteors of November 14th and of the epochs named in Chapter II. have their respective radiants, but also those of many other nights. Mr. Alexander S. Herschel, of Collingwood, England, states that fifty-six such points of divergence are now well established. We have mentioned in a previous chapter that Mr. Greg, of Manchester, has specified several epochs at which fire-balls appear, and meteoric stone-falls occur, with unusual frequency. The number of these periods will probably be increased by future observations. Perhaps the following facts may justify the designation of July 13th–14th as such an epoch:

1. On the 13th of July, 1797, a large fire-ball was seen in Göttingen.

2. On the 14th of July, 1801, a fire-ball was seen in Montgaillard.

3. On the 14th of July, 1845, a brilliant meteor was seen in London.

4. On the 13th of July, 1846, at about 9h. and 30m. P.M., a brilliant fire-ball passed over Maryland and Pennsylvania, and was seen also in Virginia, Delaware, New Jersey, New York, and Connecticut. Its course was north, about thirty degrees east, and the projection of its path on the earth's surface passed about four miles west of Lancaster, Pennsylvania, and nearly through Mauch Chunk, in Carbon County. When west of Philadelphia its angle of elevation, as seen from that city, was forty-two degrees. Consequently its altitude, when near Lancaster, was about fifty-nine miles. The projection of its visible path, on the earth's surface, was at least two hundred and fifty miles in length. Its height, when nearest Gettysburg, was about seventy miles, and it disappeared at an elevation of about eighteen miles, near the south corner of Wayne County, Pennsylvania. Its apparent diameter, as seen from York and Lancaster, was about half that of the moon, and its estimated heliocentric velocity was between twenty and twenty-five miles.

The author was assured by persons in Harford County, Maryland, and also in York, Pennsylvania, that shortly after the disappearance of the meteor a distinct report, like that of a distant cannon, was heard. As might be expected, their estimates of the interval which elapsed were different; but Daniel M. Ettinger, Esq., of York, who was paying particular attention, in expectation of a report, stated that it was a little over six minutes. This would indicate a distance of about seventy-five miles. The sound could not therefore have resulted from an explosion at or near the termination of the meteor's observed path. The inclination of the meteoric track to the surface of the earth was such that the body could not have passed out of the atmosphere. As no aerolites, however, were found beneath any part of its path, perhaps the entire mass may have been dissipated before reaching the earth.—*Silliman's Journal* for May, 1866.

5. On the 14th of July, 1847, a remarkable fall of aerolites was witnessed at Braunau, in Bohemia. Humboldt states that "the fallen masses of stone were so hot, that, after six hours, they could not be touched without causing a burn." An analysis of some of the fragments, by Fischer and Duflos, gave the following result:

Iron	91·862
Nickel	5·517
Cobalt	0·529
Copper, manganese, arsenic, calcium, magnesium, silicium, carbon, chlorine and sulphur.	2·072
	100·000

6. On the 13th of July, 1848, a brilliant fire-ball was seen at Stone-Easton, Somerset, England.

7. On the 13th of July, 1852, a large bolide was seen in London.

8. On the 14th of July, 1854, a fire-ball was seen at Senftenberg.

9. On the 13th of July, 1855, a meteor, three times as large as Jupiter, was seen at Nottingham, England.

10. "One of the most celebrated falls that have occurred of late years is that which happened on the 14th of July, 1860, between two and half-past two in the afternoon, at Dhurmsala, in India. The aerolite in question fell with a most fearful noise, and terrified the inhabitants of the district not a little. Several fragments were picked up by the natives, and carried religiously away, with the impression that they had been thrown from the summit of the Himalayas by an invisible Divinity. Lord Canning forwarded some of these stones to the British Museum and to the Vienna Museum. Mr. J. R. Saunders also sent some of the stones to Europe. It appears that, soon after their fall, the stones were *intensely cold.** They are ordinary earthy aerolites, having a specific gravity of 3·151, containing fragments of iron and iron py-

* "This is a remarkable example of a stone arriving on the earth with a temperature approaching that of the interplanetary spaces. Aerolites containing much iron, a substance which conducts heat well, get thoroughly heated by their passage through the atmosphere. But the stony aerolites, containing less iron, conducting heat badly, preserve in their interior the temperature of the locality from which they fall; their surface only is heated, and generally fused. When the stones are large, the *excessive cold* of their interior portion, which must be nearly that of interplanetary space, is remarked; but when small, they remain hot for some time."—*Dr. Phipson.*

rites; they have an uneven texture, and a pale-gray color."

11. At a quarter-past ten o'clock on the evening of July 13th, 1864, a large fire-ball was seen in New England.* The hour of its appearance, it will be observed, was nearly the same with that of the bolide of July 13th, 1846; and it is also worthy of remark that their *directions* were nearly the same. The meteor of 1864 had a tail three or four degrees in length, and the body, like that of 1846, exploded with a loud report.

12. On the 8th of July, 1186, an aerolite fell at Mons, in Belgium (Quetelet's *Physique du Globe*, p. 320). A forward motion of the node, somewhat less than that observed in the rings of November and August, would give a correspondence of dates between the falls of 1186, 1847, and 1860.

With the exception of the last, which is doubtful, these phenomena all occurred within a period of 67 years.

The Epoch of November 29.

It has been stated that in different years meteoric stones have fallen about the 29th of November. One of the most recent aerolites which can be assigned to this epoch is that which fell on the 30th of November, 1850, at Shalka, in Bengal. It may be mentioned, as at least a coincidence, that the earth passes the approximate intersection of her orbit with that of Biela's comet at the date of this epoch. Do

---

* Silliman's Journal, September, 1864.

other bodies besides the two Biela comets move in the same ellipse? It is worthy of remark that two star showers have been observed at this date: one in China, A.D. 930, the other in Europe, 1850 (see Quetelet's Catalogue). It is certainly important that the meteors of this epoch should be carefully studied.

CHAPTER V.

GEOGRAPHICAL DISTRIBUTION OF METEORIC STONES—DO AEROLITIC FALLS OCCUR MORE FREQUENTLY BY DAY THAN BY NIGHT?—DO METEORITES, BOLIDES, AND THE MATTER OF ORDINARY SHOOTING-STARS, COEXIST IN THE SAME RINGS?

PROFESSOR CHARLES UPHAM SHEPARD, of Amherst College, who has devoted special attention to the study of meteoric stones, has designated two districts of country, one in each continent, but both in the northern hemisphere, in which more than nine-tenths of all known aerolites have fallen. He remarks: "The fall of aerolites is confined principally to two zones; the one belonging to America is between 33° and 44° north latitude, and is about 25° in length. Its direction is more or less from northeast to southwest, following the general line of the Atlantic coast. Of all known occurrences of this phenomenon during the last fifty years, 92·8 per cent. have taken place within these limits, and mostly in the neighborhood of the sea. The zone of the Eastern continent —with the exception that it extends ten degrees more to the north—lies between the same degrees of latitude, and follows a similar northeast direction, but is more than twice the length of the American zone. Of all the observed falls of aerolites, 90·9 per cent. have taken place within this area, and were also concentrated in that half of the zone which extends along the Atlantic."

The facts as stated by Professor Shepard are, of course, unquestionable. It seems, however, extremely improbable that the districts specified should receive a much larger proportion of aerolites than others of equal extent. How, then, are the facts to be accounted for? We answer, the number of aerolites *seen* to fall in a country depends upon the number of its inhabitants. The ocean, deserts, and uninhabited portions of the earth's surface afford no instances of such phenomena, simply for the want of observers. In sparsely settled countries the fall of aerolites would not unfrequently escape observation; and as such bodies generally penetrate the earth to some depth, the chances of discovery, when the fall is not observed, must be exceedingly rare. Now the part of the American continent designated by Professor Shepard, it will be noticed, is the oldest and most thickly settled part of the United States; while that of the Eastern continent stretches in like manner across the most densely populated countries of Europe. This fact alone, in all probability, affords a sufficient explanation of Prof. Shepard's statement.*

Do aerolites fall more frequently by day than by night?—Mr. Alexander S. Herschel, of Collingwood, England, has with much care and industry collected and collated the known facts in regard to bolides and aerolites. One result of his investigations is that a much greater number of meteoric stones are ob-

---

* The same explanation is given by T. M. Hall, F.G.S., in the Popular Science Review for Oct. 1866.

served to fall by day than by night. From this he infers that, for the most part, the orbits in which they move are *interior* to that of the earth. The fact, however, is obviously susceptible of a very different explanation—an explanation quite similar to that of the frequent falls in particular districts. *At night the number of observers is incomparably less; and hence many aerolites escape detection.* There would seem to be no cause, reason, or antecedent probability of these falls being more frequent at one hour than another in the whole twenty-four.

The coexistence of meteorites, bolides, and the matter of shooting-stars in the same rings?—It has been stated on a previous page that several aerolite epochs are coincident with those of shooting-stars. Is the number of such cases sufficient to justify the conclusion that the correspondence of dates is not accidental? We will consider,

I. The Epoch of November 11th–14th.

1. 1548, November 6th. A very large detonating meteor was seen at Mansfield, Thuringia, at two o'clock in the morning. The known rate of movement of the node brings this meteor within the November epoch.
2. 1624, November 7th. A large fire-ball was seen at Tubingen. The motion of the node brings this also within the epoch.
3. 1765, November 11th. A bright meteoric light was observed at Frankfort.
4. 1791, November 11th. A large meteor was seen at Göttingen and Lilienthal.

5. 1803, November 13th. A fire-ball, twenty-three miles high, was seen at London and Edinburgh.

6. 1803, November 13th. A splendid meteor was seen at Dover and Harts.

7. 1808, November 11th. A fire-ball was seen in England.

8. 1818, November 13th. A fire-ball was seen at Gosport.

9. 1819, November 13th. A fire-ball was seen at St. Domingo.

10. 1820, November 12th. A large detonating meteor was seen at Cholimschk, Russia.

11. 1822, November 12th. A fire-ball appeared at Potsdam.

12. 1828, November 12th. A meteor was seen in full sunshine at Sury, France.

13. 1831, November 13th. A fire-ball was seen at Bruneck.

14. 1831, November 13th. A brilliant meteor was seen in the North of Spain.

15. 1833, November 12th. A fire-ball was seen in Germany.

16. 1833, November 13th. A meteor, two-thirds the size of the moon, was seen during the great meteoric shower in the United States.

17. 1834, November 13th. A large fire-ball was seen in North America.

18. 1835, November 13th. Several aerolites fell near Belmont, Department de l'Ain, France.

19. 1836, November 11th. An aerolitic fall occurred at Macao, Brazil.

20. 1837, November 12th. A remarkable fire-ball was seen in England.

21. 1838, November 13th. A large fire-ball was seen at Cherbourg.

22. 1849, November 13th. An extraordinary meteor appeared in Italy. "Seen in the southern sky. Varied in color; a bright cloud visible one and a half hour after; according to some a detonation heard fifteen minutes after bursting. Seen also like a stream of fire between Tunis and Tripolis, where a shower of stones fell; some of them into the town of Tripolis itself."

23. 1849, November 13th. A large meteor was seen at Mecklenburg and Breslau.

24. 1856, November 12th. A meteoric stone fell at Trenzano, Italy.

25. 1866, November 14th. At Athens, Greece, a large number of bolides was seen by Mr. J. F. Julius Schmidt, during the shower of shooting-stars. One of these fire-balls was of the first class, and left a train which was visible one hour to the naked eye.

II. The Epoch of August 7th–11th.

1. 1642, August 4th. A meteoric stone fell in Suffolk County, England.

2. 1650, August 6th. An aerolite fell in Holland. The observed motion of the node brings both these stone-falls within the epoch.

3. 1765, August 9th. A large bolide was seen at Greenwich.

4. 1773, August 8th. A fire-ball was seen at Northallerton.

5. 1800, August 8th. A large meteor was seen in different parts of North America.

6. 1802, August 10th. A fire-ball appeared at Quedlinburg.

7. 1807, August 9th. A bolide was seen at Nurenberg.

8. 1810, August 10th. A stone weighing seven and three-quarter pounds fell at Tipperary, Ireland.

9. 1816, August 7th. In Hungary a large fire-ball was seen to burst, with detonations.

10. 1817, August 7th. A brilliant fire-ball was seen at Augsburg.

11. 1818, August 10th. A meteoric stone, weighing seven pounds, fell at Slobodka, Russia.

12. 1822, August 7th. A meteorite fell at Kadonah, Agra.

13. 1822, August 7th. A large meteor was seen in Moravia.

14. 1822, August 11th. "A large mass of fire fell down with a great explosion" near Coblentz.

15. 1823, August 7th. Two meteoric stones fell in Nobleboro', Maine.

16. 1826, August 8th. A fire-ball was seen at Odensee.

17. 1826, August 11th. A bright meteor appeared at Halle.

18. 1833, August 10th. A fire-ball was seen at Worcestershire, England.

19. 1834, August 10th. A bolide appeared at Brussels.

20. 1838, August 9th. A fine meteor was seen in Germany.

21. 1839, August 7th. A splendid fire-ball was seen at sea.

22. 1840, August 7th. A bolide appeared at Naples.

23. 1841, August 10th. An aerolite fell at Iwan, Hungary.

24. 1842, August 9th. A greenish fire-ball was seen at Hamburg.

25. 1844, August 8th. A large meteor was seen in Brittany.

26. 1844, August 10th. A fire-ball was seen at Hamburg.

27. 1845, August 10th. A brilliant meteor was seen at London and Oxford.

28. 1847, August 9th. A large irregular meteor, "like a bright cloud of smoke," was seen at Brussels.

29. 1850, August 10th. A meteor as large as the moon was seen in Ireland.

30. 1850, August 10th. A very large bolide was observed in Paris.

31. 1850, August 11th. A fire-ball was seen in Paris.

32. 1853, August 7th. A bolide was observed at Glasgow.

33. 1853, August 7th. A meteor twice as large as Venus was seen at Paris.

34. 1853, August 9th. A large meteor was seen to separate into two parts.

35. 1855, August 10th. A bluish meteor, five times as large as Jupiter, was seen at Nottingham.

36. 1857, August 11th. A bolide was seen in Paris.

37. 1859, August 7th. A detonating meteor appeared in Germany.

38. 1859, August 11th. A meteoric stone fell near Albany, New York.

39. 1859, August 11th. A fine meteor was seen at Athens.

40. 1862, August 8th. A meteoric stone-fall occurred at Pillistfer, Russia.

41. 1863, August 11th. An aerolite fell at Shytal, India.

III. The Epoch of December 6th–13th.

The following falls of meteoric stones have occured at this epoch:

1. 1795, December 13th. At Wold Cottage, England.
2. 1798, December 13th. At Benares, India.
3. 1803, December 13th. At Mässing, Bavaria.
4. 1813, December 13th. At Luotolaks, Finland.
5. 1858, December 9th. At Ausson, France.
6. 1863, December 7th. At Tirlemont, Belgium.
7. 1863, December 10th. At Inly, near Trebizond.*

---

* This list contains nothing but *aerolites*. In the Edinburgh Review for January, 1867, we find the following statements: "Out of the large number of authentic aerolites preserved in mineralogical collections, two only—one on the 10th of August, and one on the 13th of November—are recorded to have fallen on star-shower dates. On the other hand, five or six meteorites, on the epoch of the 13th–14th of October, belong to a date when star-showers, so far as is at present known, do not make their appearance." The inaccuracy of the former statement is sufficiently apparent. In regard to the latter we remark that Quetelet's Catalogue gives one star-shower on the 14th of October, and another on the 12th.

IV. The Epoch of April 18th–26th.

For this epoch we have the following aerolites:

1. 1803, April 26th. At L'Aigle, France.
2. 1808, April 19th. At Casignano, Parma, Italy.
3. 1838, April 18th. At Abkurpore, India.
4. 1842, April 26th. At Milena, Croatia.

V. The Epoch of April 9th–12th.

1. 1805, April 10th. At Doroninsk, Russia.
2. 1812, April 10th. At Toulouse, France.
3. 1818, April 10th. At Zaborzika, Russia.
4. 1864, April 12th. At Nerft, Russia.

The foregoing lists, which might be extended, are sufficient to establish the fact that meteoric stones are but the largest masses in the nebulous rings from which showers of shooting stars are derived; a fact worthy of consideration whatever theory may be adopted in regard to the origin of such annuli.

CHAPTER VI.

PHENOMENA SUPPOSED TO BE METEORIC—METEORIC DUST—DARK DAYS.

It is well known that great variety has been found in the composition of aerolites. While some are extremely hard, others are of such a nature as to be easily reducible to powder. It is not impossible that when some of the latter class explode in the atmosphere they are completely pulverized, so that, reaching the earth in extremely minute particles, they are never discovered. It is very unlikely, moreover, that of the millions of shooting-stars that daily penetrate the atmosphere nothing whatever in the solid form should ever reach the earth's surface. Indeed, the celebrated Reichenbach, who devoted great attention to this subject, believed that he had actually discovered such deposits of meteoric matter. Chladni and others have detailed instances of the fall of *dust*, supposed to be meteoric, from the upper regions of the atmosphere. The following may be regarded, with more or less probability, as instances of such phenomena:

1. A.D. 475, November 5th or 6th. A shower of black dust fell in the vicinity of Constantinople. Immediately before or about the time of the fall, ac-

cording to old accounts, "the heavens appeared to be on fire," which seems to indicate a meteoric display of an extraordinary character.

2. On the 3d of December, 1586, a considerable quantity of dark-colored matter fell from the atmosphere, at Verde, in Hanover. The fall was attended by intense light, as well as by a loud report resembling thunder. The substance which fell was hot when it reached the earth, as the planks on which a portion of it was found were slightly burnt, or charred. The date of this occurrence, allowance being made for the movement of the node, is included within the limits of the meteoric epoch of December 6th–13th.

3. About a century later, viz., on the 31st of January, 1686, a very extensive deposit of blackish matter, in appearance somewhat resembling charred paper, took place in Norway and other countries in the north of Europe. A portion of this substance, which had been carefully preserved, was analyzed by Grotthus, and found to contain iron, silica, and other elements frequently met with in aerolites.

4. On the 15th of November, 1755, red rain fell in Sweden and Russia, and on the same day in Switzerland. It gave a reddish color to the waters of Lake Constance, to which it also imparted an acid taste. The rain which fell on this occasion deposited a sediment whose particles were attracted by the magnet.

5. In 1791 a luminous meteor exploded over the Atlantic Ocean, and at the same time a quantity of matter resembling sand descended to the surface.

6. According to Chladni the explosion of a large

bolide over Peru, on the 27th of August, 1792, was followed by a shower of cindery matter, the fall of which continued during three consecutive days.

7. On the 13th and 14th of March, 1813, a shower of red dust fell in Calabria, Tuscany, and Friuli. The deposit was sufficient to impart its color to the snow which was then upon the ground. That this dust was meteoric can scarcely be doubted, since at the same time a shower of aerolites fell at Cutro, in Calabria, attended by two loud reports resembling thunder. The shower of dust continued several hours, and was accompanied by a noise which was compared to the distant dashing of the waves of the ocean.*

8. In November, 1819, black rain and snow fell in Canada.

9. On the 3d of May, 1831, red rain fell near Giessen. It deposited a dark-colored sediment which Dr. Zimmermann found to contain silica, oxide of iron, and various other substances observed in aerolites.

It is well known that quantities of sand are often conveyed, by the trade-winds, from the continent of

---

* The date of this remarkable occurrence is worthy of note as a probable aerolite epoch. From the 12th to the 15th of March we have the following falls of meteoric stones:

1. 1731, March 12th. At Halstead, Essex, England.
2. 1798, March 12th. At Salés, France.
3. 1806, March 15th. At Alais, France.
4. 1807, March 13th. At Timochin, Russia.
5. 1811, March 13th. At Kuleschofka, Russia.
6. 1813, March 13th–14th. The phenomena above described.
7. 1841, March 12th. At Grüneberg, Silesia.

Numerous fire-balls have appeared at the same epoch.

Africa and deposited in the ocean. Such sand-showers have sometimes occurred several hundred miles from the coast. Volcanic matter also has been occasionally carried a considerable distance. The phenomena above described cannot, however, be referred to such causes; and there can be little doubt that most, if not all of them, were of meteoric origin.

There is, in all probability, a regular gradation from the smallest visible shooting-stars to bolides and aerolites. No doubt a great number of very small meteoric stones penetrate beneath the earth's surface and escape observation. An interesting ac count of the accidental discovery of such *celestial pebbles* has recently been given by Professor Haidinger, of Vienna. The meteor from which they were derived *was but little larger than an ordinary shooting-star.* Its track was visible, however, until it terminated at the earth's surface. Professor Haidinger's account is as follows: On the 31st of July, 1859, about half-past nine o'clock in the evening, three inhabitants of the bourg of Montpreis, in Styria, saw a small luminous globe, very similar to a shooting-star, and followed by a luminous streak in the heavens, fall directly to the earth, which it attained close to the château that exists in the locality. The fall was accompanied by a whistling or hissing noise in the air, and terminated by a *slight* detonation. The three observers, rushing to the spot where the meteor fell, immediately found a small cavity in the hard, sandy soil, from which they extracted three small meteoric stones about the size of nuts, and a quantity of black powder. For five to eight seconds these stones continued in a *state of incandescence*, and it was necessary to allow upwards of a quarter of an hour to elapse before they could be touched without inflicting a burn. They appear to have been ordinary meteoric stones, covered with the usual black rind. The possessors would not give them up to be analyzed. The details of this remarkable occurrence of the fall of an extremely small meteor, we owe to Herr Deschann, Conservator of the Museum of Laibach, in Carniola, and member of the Austrian Chamber of Deputies.

The following is perhaps the only instance on record in which a shooting-star *lower than the clouds* has been undoubtedly observed. The date is one at which meteors are said to be more than usually numerous; and the radiant point for the epoch has been recently determined, by British observers, to be about *Gamma Cygni.* The meteor was seen by Mr. David Trowbridge, of Hector, Schuyler County, New York, who says: "On the evening of July 26th, 1866, about 8h. 15m. P.M., a very bright meteor flashed out in Cygnus, and moved from east to west with great rapidity. Its path was about 30° after I saw it. Height above the northern horizon about 50°. Duration of flight from one-half to one second. It left a beautiful train. The head was red and train blue. It was certainly *below* the clouds. It passed between me and some cirro-stratus clouds, so dense as to hide ordinary stars completely. Several others that saw it said it was *below* the clouds."—*Silliman's Journal* for Sept. 1866. It seems altogether probable that when a meteor thus descends, before its explosion or dissipation, into the lower atmospheric strata, at least portions of its mass must reach the earth's surface.

Meteoric Transits—Dark Days.

If shooting-stars and aerolites are derived from meteoric rings revolving round the sun in orbits nearly intersecting that of the earth, then (1) these masses must sometimes transit the solar disk; (2) if any of the rings contain either individual masses of considerable magnitude, or sufficiently dense swarms of meteoric asteroids, such transits may sometimes be observed; (3) the passage of a dense meteoric cluster over the solar disk must partially intercept the sun's light and heat; and (4) should both nodes of the ring very nearly intersect the earth's orbit, meteoric falls might occur when the earth is at either; in which case the epochs would be separated by an interval of about six months. Have any such phenomena as those indicated been actually observed?

The passage of dark spots across the sun, having a much more rapid motion than the solar maculæ, has been frequently noticed. The following instances are well authenticated:

1779, June 17th. About mid-day the eminent French astronomer, Messier, saw a great number of black points crossing the sun. Rapidly moving spots were also seen by Pastorff on the following dates:

1822, October 23d,

1823, July 24th and 25th,

1836, October 18th,

and on several subsequent occasions the same astronomer witnessed similar phenomena. Another transit of this kind has been seen quite recently. On the 8th of May, 1865, a small black spot was seen by Coumbary to cross the solar disk. It seems difficult to account for these appearances (so frequently seen by experienced observers) unless we regard them as meteoric masses.

Partial Interception of the Sun's Light and Heat.

Numerous instances are on record of partial obscurations of the sun which could not be accounted for by any known cause. Cases of such phenomena took place, according to Humboldt, in the years 1090, 1203, and 1547. Another so-called *dark day* occurred on the 12th of May, 1706, and several more (some of still later date) might be specified. Chladni and other physicists have regarded the transit of meteoric masses as the most probable cause of these obscurations. It is proper to remark, however, that the eminent French astronomer, Faye, who has given the subject much attention, finds little or no evidence in support of this conjecture.

An examination of meteorological records is said to have established two epochs of abnormal cold, viz., about the 12th of February and the 12th of May. The former was pointed out by Brandes about the beginning of the present century; the latter by Mädler, in 1834. The May epoch occurs when the earth is in conjunction with one of the nodes of the November meteoric ring; and that of February has a similar relation to the August meteors. M. Erman, a distinguished German scientist, soon after the discovery of the August and November meteoric epochs, suggested that those depressions of temperature might be explained by the intervention of the meteoric zones between the earth and the sun. The period, however, of the November meteors being still somewhat doubtful, their position with respect to the earth about the 12th of May is also uncertain. But however this may be, the following dates of aerolitic falls seem to indicate May 8th–14th, or especially May 12th–13th, as a meteoric epoch:

(*a*) May 8th, 1829, Forsyth, Georgia, U. S. A.

(*b*) May 8th, 1846, Macerata, Italy.

(*c*) May 9th, 1827, Nashville, Tennessee, U. S. A.

(*d*) May 12th, 1861, Goruckpore, India.

(*e*) May 13th, 1831, Vouillé, France.

(*f*) May 13th, 1855, Oesel, Baltic Sea.

(*g*) May 13th, 1855, Bremevörde, Hanover.

(*h*) May 14th, 1861, near Villanova, in Catalonia, Spain.

(*i*) May 14th, 1864, Orgueil, France.

All the foregoing, except that of May 14th, 1861, may be found in Shepard's list, *Silliman's Journal* for January, 1867.

It has been shown in a former chapter that more than seven millions of shooting-stars of sufficient magnitude to be seen by the naked eye daily enter the earth's atmosphere. As the small ones are the most numerous, it is not improbable that an indefinitely greater number of meteoric particles, too minute to be visible, are being constantly, in this manner, arrested in their orbital motion. Now, it would certainly be a very unwarranted conclusion that these atmospheric increments are all of a permanently gaseous form. In view of this strong probability that meteoric dust is daily reaching the earth's surface, Baron von Reichenbach, of Vienna, conceived the idea of attempting its discovery. Ascending to the tops of some of the German mountains, he carefully collected small quantities of the soil from positions in which it had not been disturbed by man. This matter, on being analyzed, was found to contain small portions of nickel and cobalt—elements rarely found in the mineral masses scattered over the earth's surface, but very frequently met with in aerolites. In short, Reichenbach believed, and certainly not without some probability, that he had detected minute portions of meteoric matter.

CHAPTER VII.

FURTHER RESEARCHES OF REICHENBACH—THEORY OF METEORS—STABILITY OF THE SOLAR SYSTEM—DOCTRINE OF A RESISTING MEDIUM.

THE able and original researches of the celebrated Reichenbach, who has made meteoric phenomena the subject of long-continued and enthusiastic investigation, have attracted the general attention of scientific men. It is proposed to present, in the following chapter, a brief *resumé* of his views and conclusions.

1. *The Constitution of Comets.*—It is a remarkable fact that cometary matter has no refractive power, as is manifest from the observations of stars seen through their substance.* These bodies, therefore, are not gaseous; and the most probable theory in regard to their nature is that they consist of an infinite number of discrete, solid molecules, at great distances from each other, with very little attraction among themselves, or toward the nucleus, and having, therefore, great mobility. Now Baron Reichenbach, having carefully examined a great number of meteoric stones, has found them for the most part

* The innermost or semi-transparent ring of Saturn appears to be similarly constituted, as the body of the planet is seen through it without any distortion whatever.

composed of extremely minute globules, apparently cemented together. He hence infers that they have been comets—perhaps very small ones—whose component molecules have by degrees collected into single masses.

2. *The Number of Aerolites.*—The average number of aerolitic falls in a year was estimated by Schreibers, as previously stated, at 700. Reichenbach, however, after a thorough discussion of the data at hand, makes the number much larger. He regards the probable annual average, for the entire surface of the earth, as not less than 4500. This would give about twelve daily falls. They are of every variety as to magnitude, from a weight of less than a single ounce to over 30,000 pounds. The Baron even suspects the meteoric origin of large masses of dolerite which all former geologists had considered native to our planet. In view of the fact that from the largest members of our planetary system down to the particles of meteoric dust there is an approximately regular gradation, and that the larger, at least in some instances, appear to have been formed by the aggregation of the smaller, he asks may not the earth itself have been formed by an agglomeration of meteorites? The learned author, from the general scope of his speculations, would thus seem to have adopted a form of the nebular hypothesis somewhat different from that proposed by Laplace.

3. *Composition and mean Density of Aerolites.*—A large proportion of meteoric stones are similar in structure to the volcanic or plutonic rocks of the earth; and *all* consist of elements identical with those in our planet's crust. Their mean density,

moreover, is very nearly the same with that of the earth. These facts are regarded by Reichenbach as indicating that those meteoric masses which are daily becoming incorporated with our planet, have had a common origin with the earth itself. Baron Reichenbach's views, as presented by himself, will be found at length in *Poggendorf's Annalen* for December, 1858.

Stability of the Solar System.—The well-known demonstrations of the stability of the solar system, given by Lagrange and Laplace, are not to be accepted in an unlimited sense. They make no provision against the destructive agency of a resisting medium, or the entrance of matter into the solar domain from the interstellar spaces. In short, the conservative influence ascribed to these celebrated theorems extends only to the major planets; and even in their case it is to be understood as applying only to their mutual perturbations. The phenomena of shooting-stars and aerolites have demonstrated the existence of considerable quantities of matter moving in *unstable* orbits. The amount of such matter within the solar system cannot now be determined; but the term probably includes the zodiacal light, many, if not all, of the meteoric rings, and a large number of comets. These unstable parts of the system are being gradually incorporated with the sun, the earth, and doubtless also with the other large planets. It is highly probable that at former epochs the quantity of such matter was much greater than at present, and that, unless new supplies be received *ab extra*, it must, by slow degrees, disappear from the system.

The fact, now well established, of the extensive diffusion of meteoric matter through the interplanetary spaces has an obvious bearing on Encke's theory of a resisting medium. If we grant the existence of such an ether, it would seem unphilosophical to ascribe to it one of the properties of a material fluid—the power of resisting the motion of all bodies moving through it—and to deny it such properties in other respects. Its condensation, therefore, about the sun and other large bodies must be a necessary consequence. This condensation existed in the primitive solar spheroid, before the formation of the planets: the rotation of the spheroid would be communicated to the coexisting ether; and hence, *during the entire history of the planetary system, the ether has revolved around the sun in the same direction with the planets.* This condensed ether, it is also obvious, must participate in the progressive motion of the solar system.

But again; even if we reject the doctrine of the development of the planetary bodies from a rotating nebula, we must still regard the density of the ether as increasing to the center of the system. The sun's rotation, therefore, would communicate motion to the first and denser portions; this motion would be transmitted outward through successive strata, with a constantly diminishing angular velocity. The motion of the planets themselves through the medium in nearly circular orbits would concur in imparting to it a revolution in the same direction. Whether, therefore, we receive or reject the nebular hypothesis, the resistance of the ethereal medium to bodies moving in orbits of small eccentricity and in the direction of the sun's rotation, becomes an infinitesimal quantity.

The hypothesis of Encke, it is well known, was based solely on the observed acceleration of the comet which bears his name. More recently, however, a still greater acceleration has been found in the case of Faye's comet. Now as the meteoric matter of the solar system is a *known* cause for such phenomena, sufficient, in all probability, both in mode and measure, the doctrine of a resisting ethereal medium would seem to be a wholly unnecessary assumption.

CHAPTER VIII.

DOES THE NUMBER OF AEROLITIC FALLS VARY WITH THE EARTH'S DISTANCE FROM THE SUN?—RELATIVE NUMBERS OBSERVED IN THE FORENOON AND AFTERNOON—EXTENT OF THE ATMOSPHERE AS INDICATED BY METEORS.

An analysis of any extensive table of meteorites and fire-balls proves that a greater number of aerolitic falls have been observed during the months of June and July, when the earth is near its aphelion, than in December and January, when near its perihelion. It is found, however, that the reverse is true in regard to bolides, or fire-balls. Now the theory has been held by more than one physicist, that aerolites are the outriders of the asteroid ring between Mars and Jupiter; their orbits having become so eccentric that in perihelion they approach very near that of the earth. If this theory be the true one, the earth would probably encounter the greatest number of those meteor-asteroids when near its aphelion. The hypothesis therefore, it has been claimed, appears to be supported by well-known facts. The variation, however, in the observed number of aerolites may be readily accounted for independently of any theory as to their origin. The fall of meteoric stones would evidently be more likely to escape observation by night than by day, by reason of the relatively small number of observers. But the days are shortest when the earth is in perihelion, and longest when in aphelion; the ratio of their lengths being nearly equal to that of the corresponding numbers of aerolitic falls.

On the other hand, it is obvious that fire-balls, unless of very extraordinary magnitude, would not be visible during the day. The *observed* number will therefore be greatest when the nights are longest; that is, when the earth is near its perihelion. This, it will be found, is precisely in accordance with observation.

It has been found, moreover, that a greater number of meteoric stones fall during the first half of the day, that is, from midnight to noon, than in the latter half, from noon to midnight. This would seem to indicate that a large proportion of the aerolites encountered by the earth have direct motion.

Height of the Atmosphere.—The weight of a given volume of mercury is 10,517 times that of an equal volume of air at the earth's surface; and since the mean height of the mercurial column in the barometer is about thirty inches, if the atmosphere were of uniform density its altitude would be about 26,300 feet, or nearly five miles. The density rapidly diminishes, however, as we ascend above the earth's surface. Calling it unity at the sea level, the rate of variation is approximately expressed as follows:

Altitude in Miles.	Density.
0	1
7	$\frac{1}{4}$
14	$\frac{1}{16}$
21	$\frac{1}{64}$
28	$\frac{1}{256}$
35	$\frac{1}{1024}$
70	$\frac{1}{1000000}$
105	$\frac{1}{1000000000}$
140	$\frac{1}{1000000000000}$
etc.	etc.

From this table it will be seen that at the height of 35 miles the air is one thousand times rarer than at the surface of the earth; and that, supposing the same rate of decrease to continue, at the height of 140 miles the rarity would be one trillion times greater. The atmosphere, however, is not unlimited. When it becomes so rare that the force of repulsion between its particles is counterbalanced by the earth's attraction, no further expansion is possible. To determine the altitude of its superior surface is a problem at once difficult and interesting. Not many years since about 45 or 50 miles were generally regarded as a probable limit. Considerable light, however, has been thrown upon the question by recent observations in meteoric astronomy. Several hundred detonating meteors have been observed, and their average height at the instant of their first appearance has been found to exceed 90 miles. The great meteor of February 3d, 1856, seen at Brussels, Geneva, Paris, and elsewhere, was 150 miles high when first seen, and a few apparently well-authenticated instances are known of a still greater elevation. We conclude, therefore, from the evidence afforded by meteoric phenomena, that the height of the atmosphere is certainly not *less* than 200 miles.

It might be supposed, however, that the resistance of the air at such altitudes would not develop a sufficient amount of heat to give meteorites their brilliant appearance. This question has been discussed by Joule, Thomson, Haidinger, and Reichenbach, and may now be regarded as definitively settled. When the velocity of a meteorite is known the quantity of heat produced by its motion through air of a given density is readily determined. The temperature acquired is the equivalent of the force with which the atmospheric molecules are met by the moving body. This is about one degree (Fahrenheit) for a velocity of 100 feet per second, and it varies directly as the square of the velocity. A velocity, therefore, of 30 miles in a second would produce a temperature of 2,500,000°. The weight of 5280 cubic feet of air at the earth's surface is about 2,830,000 grains. This, consequently, is the weight of a column 1 mile in length, and whose base or cross section is one square foot. The weight of a column of the same dimensions at a height of 140 miles would be about $\frac{1}{350000}$th of a grain. Hence the heat acquired by a meteoric mass whose cross section is one square foot, in moving 1 mile would be one grain raised $7\frac{1}{7}$ degrees, or one-fifth of a grain 2500° in 70 miles. This temperature would undoubtedly be sufficient to render meteoric bodies brilliantly luminous.

But there have been indications of an atmosphere at an elevation of more than 500 miles. A discussion of the best observations of the great aurora seen throughout the United States on the 28th of August, 1859, gave 534 miles as the height of the upper limit above the earth's surface. The aurora of September 2d, of the same year, had an elevation but little inferior, viz., 495 miles. Now, according to the observed rate of variation of density, at the height of 525 miles, the atmosphere would be so rare that a sphere of it filling the orbit of Neptune would contain less matter than $\frac{1}{30}$th of a cubic inch of air at the earth's surface. In other words, it would weigh less than $\frac{1}{90}$th of a grain. We are thus forced to the conclusion either that the law of variation is not the same at great heights as near the surface; or, that beyond the limits of the atmosphere of air, there is another of electricity, or of some other fluid.

CHAPTER IX.

THE METEORIC THEORY OF SOLAR HEAT.

Of the various theories proposed by astronomers to account for the origin of the sun's light and heat, only two have at present any considerable number of advocates. These are—

1. *The Chemical Theory;* according to which the light and heat of the sun are produced by the chemical combination of its elements; in other words, by an intense combustion.

2. *The Meteoric Theory*, which ascribes the heat of our central luminary to the fall of meteors upon its surface. The former is advocated with great ingenuity by Professor Ennis in a recent work on "*The Origin of the Stars, and the Causes of their Motions and their Light.*" It has, on the other hand, been ably opposed by Dr. Mayer, Professor William Thomson, and other eminent physicists. A brief examination of its claims may not be destitute of interest.

If the sun's heat is produced by chemical action, whence comes the necessary supply of fuel to support the combustion? The quantity of solar heat radiated into space has been determined with at least an approximation to mathematical precision. We know also the amount produced by the combustion of a given quantity of coal. Now it has been found by calculation that if the sun were a solid globe of coal, and a sufficient supply of oxygen were furnished to support its combustion, the amount of heat resulting from its consumption would be less than that actually emitted during the last 6000 years. In short, *no known* elements would meet the demands of the case. But it is highly probable that the different bodies of the solar system are composed of the same elements. This view is sustained by the well-known fact that meteoric stones, which have reached us from different and distant regions of space, have brought us no new elementary substances. The *chemical* theory of solar heat seems thus encumbered with difficulties well-nigh insuperable.

Professor Ennis' mode of obviating this objection, though highly ingenious, is by no means conclusive. The latest analyses of the solar spectrum indicate, he affirms, the presence of numerous elements besides those with which we are acquainted. Some of these may yield by their combustion a much greater amount of heat than the same quantity of any known elements in the earth's crust. "Every star," he remarks, "as far as yet known, has a different set of fixed lines, although there are certain resemblances between them. They lead to the conclusion that each star has, in part at least, its peculiar modifications of matter, called simple elements; but the number of stars is infinite, and therefore the number of elements must be infinite."* He argues, moreover, that in a globe so vast as the sun there may be forces in operation with whose nature we

---

* Origin of the Stars, p. 173.

are wholly unacquainted. This leaving of the *known* elements as well as the *known* laws of nature for *unknown possibilities* will hardly be satisfactory to unbiased minds.

Again: that the different bodies of the universe are composed of different elements is inferred by our author from the following among other considerations: "In our solar system Mercury is sixty or eighty times more dense than one of the satellites of Jupiter, and probably in a much greater proportion denser than the satellites of Saturn. This indicates a wide difference between the nature of their elements." This statement is again repeated in a subsequent page.* "The densities of the planets and their satellites prove that they are composed of very different elements. Mercury is more than sixty times, and our earth about fifty times, more dense than the inner moon of Jupiter. Saturn is only about one-ninth as dense as the earth; it would float buoyantly on water. There is a high probability that the satellites of Saturn and Uranus are far lighter than those of Jupiter. Between the two extremes of the attendants of the sun, there is probably a greater difference in density than one hundred to one; and from one extreme to the other there are regular gradations of small amount.

"The difference in constitution between the earth and the moon is seen in their densities: that of the moon being about half that of the earth. The nitrogen of our globe is found only in the atmosphere, and such substances as derive it from the

---

* Origin of the Stars, p. 184.

atmosphere. The moon has no appreciable atmosphere, and therefore, in a high probability, no nitrogen."

The statements here quoted were designed to show that the physical constitution of the sun and planets is widely different from that of the earth, and that the combustion of *some* of the elements in this indefinite variety may account for the origin of solar heat. Let us examine the facts.

According to Laplace the mass of Jupiter's first satellite is 0·000017328, that of Jupiter being 1. The diameter is 2436 miles. Hence the corresponding density is a little more than *one-fifth* of the mean density of the earth. In other words, it is somewhat greater than the density of water, and very nearly equal to that of Jupiter himself. Professor Ennis' value is therefore erroneous.* In regard to the densities of the Saturnian and Uranian satellites nothing is known, and conjecture is useless. In short, Saturn has the least mean density of all the planets, primary or secondary, so far as known. This may be owing to the great extent of his atmospheric envelope. The density of the moon is but three-fifths that of the earth: it is to be borne in mind, however, that the *mass* and *pressure* are also much less.

With respect to meteorites the same author remarks that "like the moon, they are probably satellites of the earth; but being very small, they are

---

* Since the above was written Prof. Ennis has informed the author that, without making any estimate of his own, he adopted the density of Jupiter's first satellite as given in Lardner's *Handbook of Astronomy*.

liable to extraordinary perturbations, and hence strike the earth in many directions." Here, again, his *facts* are at fault; for (1) the observed velocities of these bodies are inconsistent with the supposition of their being satellites of the earth; and (2) the amount of perturbation of such bodies does not vary with their masses: a *small* meteorite would fall toward the earth or any other planet with no greater velocity than a *large* one.

The Meteoric Theory.

It has been shown in a previous chapter that immense numbers of meteoric asteroids are constantly traversing the planetary spaces—that many millions, in fact, daily enter the earth's atmosphere. Reasons are not wanting for supposing the numbers of these bodies to increase with great rapidity as we approach the center of the system. Moreover, on account of the greater force of gravity at the sun's surface the heat produced by their fall must be much greater than at the surface of the earth. It has been calculated that if one of these asteroids be arrested in perihelion by the solar atmosphere, the quantity of heat thus developed will be 9000 times greater than that produced by the combustion of an equal mass of coal. There can, therefore, be no reasonable doubt that a *portion* of the sun's heat is produced by the impact of meteoric matter. In considering the probability that it is *chiefly* so generated, the following questions naturally present themselves:

1. *What amount of matter precipitated upon the sun would develop the quantity of heat actually emitted?*—

This question has been satisfactorily discussed by eminent physicists, and it will be sufficient for our purpose to give the result. According to Professor William Thomson, of Glasgow, the present rate of emission would be kept up by a meteoric deposit which would form an annual stratum 60 feet in thickness over the sun's surface.

2. *Could such an increase of the sun's magnitude be detected by micrometrical measurement?*—This inquiry is readily answered in the negative. The apparent diameter would be augmented only one second in 17,600 years.

3. *Is there any known or visible source from which this amount of meteoric matter may be supplied?*—Thomson, Mayer, and other distinguished writers regard the zodiacal light as the source of such meteorites. The inner portions of this immense "tornado" must be resisted in their motions by the solar atmosphere, and hence precipitated upon the sun's surface.

4. *Would this increase of the sun's mass derange the motions of the solar system?*—To this question Prof. Ennis gives an affirmative answer; his first objection to the theory under consideration being stated as follows: "The constant accumulation of such materials, during hundreds of millions of years, would increase the body of the sun and its consequent gravity so greatly as to derange the entire solar system, by destroying the balance between the centripetal and centrifugal forces now acting on the planets."* This, it must be confessed, would be a valid objection, if the meteoric matter were sup-

* Origin of the Stars, p. 77.

posed to be derived from the extra-planetary spaces. As their source, however,—the zodiacal light—is interior to the earth's orbit, it can have no application to any planet exterior to Venus. Most probably the greater portion of the meteoric mass is even within the orbit of Mercury, so that the effect of its convergence could scarcely be noticed even in the motion of the interior planets. In pre-historic time the zodiacal light may have extended far beyond the earth's orbit. If so, its convergence to its present dimensions was undoubtedly attended by an acceleration of the earth's mean motion. We can of course have no evidence that such a shortening of the year has never occurred.

The second objection urged against the meteoric theory by the author of "The Origin of the Stars" is thus expressed: "As we must believe that all stars were lighted up by the same means, so we must believe, according to this theory, that the present interior heat of the earth and its former melted condition in both exterior and interior, was caused by the fall of meteorites. But if so, they must have gradually ceased to fall, as space became cleared of their presence, and we would now find a thick covering of meteorites on the earth's cooled surface. Instead of this, we find them very rarely, and in accordance with their present very rare falls."

To this it may be replied that the primitive igneous fluidity of the earth and planets was a necessary consequence of their condensation—a fact which has no inconsistency with the theory in question.

A different *mechanical* theory of the origin of solar heat is advocated by Professor Helmholtz in his interesting work *On the Interaction of Natural Forces.* In regard to the sun he says: "If we adopt the very probable view, that the remarkably small density of so large a body is caused by its high temperature, and may become greater in time, it may be calculated that if the diameter of the sun were diminished only the ten-thousandth part of its present length, by this act a sufficient quantity of heat would be generated to cover the total emission for 2100 years. Such a small change besides it would be difficult to detect by the finest astronomical observations."* The same view is adopted by Dr. Joel E. Hendricks, of Des Moines, Iowa.†

* Youman's Correlation and Conservation of Forces, p. 244.

† Iowa Instructor and School Journal for November, 1866, p. 49.

CHAPTER X.

WILL THE METEORIC THEORY ACCOUNT FOR THE PHENOMENA OF VARIABLE AND TEMPORARY STARS?

Having shown that meteor-asteroids are diffused in vast quantities throughout the universe; that according to eminent physicists the solar heat is produced by the precipitation of such matter on the sun's surface; and that Leverrier has found it necessary to introduce the disturbing effect of meteoric rings in order fully to account for the motion of Mercury's perihelion; we now propose extending the meteoric theory to a number of phenomena that have hitherto received no satisfactory explanation.

Variable and Temporary Stars.

No theory as to the origin of the sun's light and heat would seem to be admissible unless applicable also to the sidereal systems. Will the meteoric theory explain the phenomena of variable and temporary stars?

"It has been remarked respecting variable stars, that in passing through their successive phases, they are subject to sensible irregularities, which have not hitherto been reduced to fixed laws In general they do not always attain the same maximum brightness, their fluctuations being in some cases very considerable. Thus, according to Argelander, the variable star in *Corona Borealis*, which Pigott discovered in 1795, exhibits on some occasions such feeble changes of brightness, that it is almost impossible to distinguish the maxima from the minima by the naked eye; but after it has completed several of its cycles in this manner, its fluctuations all at once become so considerable, that in some instances it totally disappears. It has been found, moreover, that the light of variable stars does not increase and diminish symmetrically on each side of the maximum, nor are the successive intervals between the maxima exactly equal to each other."—*Grant's History of Physical Astronomy*, p 541.

Of the numerous hypotheses hitherto proposed to account for these phenomena we believe none can be found to include and harmonize all the facts of observation. The theories of Herschel and Maupertius fail to explain the irregularity in some of the periods; while those of Newton and Dunn afford no explanation of the periodicity itself.* But let us suppose that among the fixed stars some have atmospheres of great extent, as was probably the case with the sun at a remote epoch in its history. Let us also suppose the existence of nebulous rings, like those of our own system, moving in orbits so elliptical

* A recent hypothesis in regard to the temporary star of 1572 has been proposed by Alexander Wilcocks, M.D., of Philadelphia. See Journ. Acad. Nat. Sci. of Phila. for 1859.

that in their perihelia they pass through the atmospheric envelopes of the central stars. Such meteoric rings of varying density, like those revolving about the sun, would evidently produce the phenomena of variable stars. The resisting medium through which they pass in perihelion must gradually contract their orbits, or, in other words, diminish the intervals between consecutive maxima. Such a shortening of the period is now well established in the case of *Algol*. Again, if a ring be influenced by perturbation the period will be variable, like that of *Mira Ceti*. A change, moreover, in the perihelion distance will account for the occasional increase or diminution of the apparent magnitude at the different maxima of the same star. But how are we to account for the variations of brightness observed in a number of stars where no order or periodicity in the variation has as yet been discovered? It is easy to perceive that either a single nebulous ring with more than one *hiatus*, or several rings about the same star, may produce phenomena of the character described. Finally, if the matter of an elliptic ring should accumulate in a single mass, so as to occupy a comparatively small arc, its passage through perihelion might produce the phenomenon of a so-called temporary star.

Recent researches relating to nebulæ seem in some measure confirmatory of the view here presented. These observations have shown (1) a change of position in some of these objects, rendering it probable that in certain cases they are not more distant than fixed stars visible to the naked eye; and (2) a variation in the brilliancy of many small stars situated in the great nebula of Orion, and also the existence of numerous masses of nebulous matter in the form of tufts apparently attached to stars,—facts regarded as indicative of a physical connection between the stars and nebulæ.*

* Gautier's Notice of Recent Researches relating to Nebulæ.—Silliman's Journal for Jan. 1863, and March, 1864.

CHAPTER XI.

THE LUNAR AND SOLAR THEORIES OF THE ORIGIN OF AEROLITES.

Besides the *cosmical* theory of aerolites which has been adopted in this work, and which is now accepted by a great majority of scientific men, at least four others have been proposed: (1) the *atmospheric*, according to which they are formed, like hail, in the earth's atmosphere; (2) the *volcanic*, which regards them as matter ejected with great force from terrestrial volcanoes; (3) the *lunar*, which supposes them to have been thrown from craters in the moon; and (4) the *solar* hypothesis, according to which they are projected by some tremendous explosive force from the great central orb of our system. The first and second have been universally abandoned as untenable. The third and fourth, however, are entitled to consideration.

The Lunar Theory.

The theory which regards meteoric stones as products of eruption in lunar volcanoes was received with favor by the celebrated Laplace: "As the gravity at the surface of the moon," he remarks, "is much less than at the surface of the earth, and as this body has no atmosphere which can oppose a sensible resistance to the motion of projectiles, we may conceive that a body projected with a great force, by the explosion of a lunar volcano, may attain and pass the limit, where the attraction of the earth commences to predominate over that of the moon. For this purpose it is sufficient that its initial velocity in the direction of the vertical may be 2500 meters in a second; then in place of falling back on the moon, it becomes a satellite of the earth, and describes about it an orbit more or less elongated. The direction of its primitive impulsion may be such as to make it move directly toward the atmosphere of the earth; or it may not attain it, till after several and even a great number of revolutions; for it is evident that the action of the sun, which changes in a sensible manner the distances of the moon from the earth, ought to produce in the radius vector of a satellite which moves in a very eccentric orbit, much more considerable variations, and thus at length so diminish the perigean distance of the satellite, as to make it penetrate our atmosphere. This body traversing it with a very great velocity, and experiencing a very sensible resistance, might at length precipitate itself on the earth; the friction of the air against its surface would be sufficient to inflame it, and make it detonate, provided that it contained ingredients proper to produce these effects, and then it would present to us all those phenomena which meteoric stones exhibit. If it was satisfactorily proved that they are not produced by volcanoes, or generated in our atmosphere, and that their cause must be sought beyond it, in the regions of the heavens, the preceding hypothesis, which likewise explains the identity of composition observed in meteoric stones, by an identity of origin, will not be devoid of probability."—*Système du Monde*, t. ii. cap. v.

Knowing the masses and volumes of the earth and moon, it is easy to estimate the force of gravity at their surfaces, the distance from each to the point of equal attraction, and the force with which a projectile must be thrown from the lunar surface to pass within the sphere of the earth's influence. It has been calculated that an initial velocity of about a mile and a half in a second would be sufficient for this purpose—a force not greater than that known to have been exerted by terrestrial volcanoes. The *possibility*, therefore, that volcanic matter from our satellite may reach the earth's surface seems fairly admissible.

Since the time of Laplace, several distinguished European astronomers have regarded the lunar hypothesis as more or less probable. It was advocated as recently as 1851 by the late Prof. J. P. Nichol, of Glasgow. This popular and interesting writer, after describing Tycho, a large and well-known lunar crater, from which luminous rays or stripes radiate over a considerable part of the moon's surface, expresses the opinion that that immense cavity was formed by a single tremendous explosion. "Reflecting," he remarks, "on the probable suddenness and magnitude of that force, or rather of that *explosive* energy one of whose acts we have traced, as well as on the immense mass of matter which seems to have been thus violently dispersed, is not the inquiry a natural one, *where is that matter now?* It is a mass indeed which cannot well have wholly disappeared. It filled a cavern 55 miles in breadth, and 17,000 feet deep—a cavern into which even now one might cast Chimborazo and Mont Blanc, and room be left for Teneriffe behind! Like rocks flung aloft by our volcanoes, did this immense mass fall back in fragments to the surface of the moon, or was the expulsive force strong enough to give it an outward velocity sufficient to resist the attractive power of its parent globe? The moon, be it recollected, is very small in *mass* compared with the earth, and her attractive energy greatly inferior accordingly. Laplace has even calculated that the force urging a cannon-ball, increased to a degree quite within the limits of what is conceivable, could effect a final separation between our satellite and any of its component parts. It is *possible* then, and, although not demonstrable, very far from a chimera, that the disrupted and expelled masses were, in the case of which we are speaking, driven conclusively into space; but if so, where are they now? where their new residence, and what their functions? In the emergency to which I refer, such fragments would necessarily wander among the inter-planetary spaces in most irregular orbits, and chiefly in the neighborhood of the moon and the earth. Now, while the planetary orbits are so nicely adjusted that neither confusion nor interference can ever occur, it is not at all likely that the same order could be established here; nay, it is next to certain, that in the course of its orbital revolution our globe would ever and anon come in contact with these lunar fragments; in other words, STONES *would fall occasionally to its surface, and ap-*

parently from its atmosphere."—*Planetary System*, pp. 301, 302.

We have preferred to give the views of these eminent scientists in their own language. Olbers, Biot, and Poisson, who adopted the same theory, estimated the *initial* velocity which would be necessary in order that lunar fragments might pass the point of equal attraction, and also the *final*, or acquired velocity on reaching the earth's surface. The several determinations of the former were as follows:

According to	Olbers	1·570	miles a second.
"	Biot	1·569	" "
"	Laplace	1·483	" "
"	Poisson	1·437	" "

The mean being almost exactly a mile and a half. The velocity on reaching our planet, according to Olbers, would be about six and a half miles. At the date of these calculations, however, the true velocity of aerolites had not been in any case satisfactorily determined. Since that time it has been found in numerous instances to exceed *twenty miles a second*—a velocity greater than that of the earth's orbital motion. This fact of itself would seem fatal to the theory of a lunar origin.

At the meeting of the American Association for the Advancement of Science, in 1859, Dr. B. A. Gould read a paper on the supposed lunar origin of aerolites, in which the hypothesis was subjected to the test of a rigid mathematical analysis. We will not attempt even an abstract of this interesting memoir. It amounts, however, to a virtual disproof of the lunar hypothesis.

THE SOLAR THEORY.

The theory which ascribes a solar origin to meteorites is not of recent date, having been held by Diogenes Laertius and other ancient Greeks. Among the moderns its advocates have been much less numerous than those of the lunar hypothesis. The late Professor Charles W. Hackley, of New York, regarded shooting-stars, aerolites, and even comets, as matter projected with enormous force from the solar surface. The corona seen during total eclipses of the sun he supposed to be the emanations of this matter through the intervals of the luculi.—(See the Proceedings of the American Association for the Advancement of Science, Fourteenth Meeting, 1860.) An ingenious theory, differing in its details from that of Professor Hackley, though somewhat similar in its general features, has lately been advocated by Alexander Wilcocks, M.D., of Philadelphia, in a memoir read before the American Philosophical Society, May 20th, 1864, and published in their Proceedings. In regard to this hypothesis it seems sufficient to remark that it fails to give a satisfactory account of the annual periodicity of meteoric phenomena.

CHAPTER XII.

THE RINGS OF SATURN.

UNTIL about the middle of the present century the rings of Saturn were universally regarded as solid and continuous. The labors, however, of Professors Bond and Pierce, of Cambridge, Massachusetts, as well as the more recent investigations of Prof. Maxwell, of England, have shown this hypothesis to be wholly untenable. The most probable opinion, based on the researches of these astronomers, is, that they consist of streams or clouds of meteoric asteroids. The zodiacal light and the zone of small planets between Mars and Jupiter appear to constitute analogous *primary* rings. In the latter, however, a large proportion of the primitive matter seems to have collected in distinct, segregated masses. These meteoric zones have probably presented—what are not elsewhere found in the solar system—cases of commensurability in the planetary periods. The interior satellites of Saturn are so near the ring as doubtless to exert great perturbative influence. Unfortunately, the elements of the Saturnian system as determined by different astronomers are somewhat discordant. This, however, is by no means surprising when we consider the great distance of the planet and the small magnitude of some of the satellites. For convenience of reference the mean apparent distances of the satellites, together with their periodic times, are given in the following table. The former are taken from Hind's *Solar System;* the latter from Herschel's *Outlines of Astronomy.*

TABLE I.—The Satellites of Saturn.

Name.	Sidereal Revolution.				Mean Apparent Distance.
	d.	*h.*	*m.*	*s.*	″
Mimas	0	22	37	22·9	26·78
Enceladus	1	8	53	6·7	34·38
Tethys	1	21	18	25·7	42·57
Dione	2	17	41	8·9	54·54
Rhea	4	12	25	10·8	76·16
Titan	15	22	41	25·2	176·55
Hyperion	22	12?	...	......	213·3?
Japetus	79	7	53	40·4	514·52

The late Professor Bessel devoted much attention to the theory of Titan, whose mean distance he found to be 20·706 equatorial radii of the primary. Struve's measurements of the ring are given in the second column of the following table. Sir John Herschel, however, regards the Russian astronomer's interval between the rings as "somewhat too small."* This remark is confirmed by the measurements of Encke, whose results are given in column third. The fourth contains the *mean* of Struve's and Encke's measurements; and the fifth, the same, expressed in equatorial radii of Saturn.

* Outlines of Astronomy, Art. 442.

TABLE II.—THE RINGS OF SATURN.

	STRUVE.	ENCKE.	MEAN.	IN SEMI-DIAM. OF SATURN.
	″	″	″	
Equatorial radius of the planet............	8·9955			
Ext. semi-diameter of exterior ring.........	20·047	20·2225	20·13475	2·23830
Int. semi-diameter of exterior ring.........	17·644	18·0190	17·83150	1·98230
Ext. semi-diameter of interior ring..........	17·237	17·3745	17·30575	1·92380
Int. semi diameter of interior ring..........	13·334	13·3780	13·35600	1·48470
Breadth of interval...	00·407	00·6445	00·52575	0·05844

The period of a satellite revolving at the distance, 1·9238, the interior limit of the interval		=10h.	50m.	16s.
One-sixth of the period of	Dione .	=10	56	53
One-third "	Enceladus.	=10	59	22
One-half "	Mimas .	=11	18	32
One-fourth "	Tethys .	=11	19	36
And the period of a satellite at the distance, 1·9823, the exterior limit of the interval		=11	28	3

The interval, therefore, occupies precisely the space in which the periods would be commensurable with those of the four members of the system immediately exterior. Particles occupying this portion of the *primitive* ring would always come into conjunction with one of these satellites in the same parts of their orbits. Such orbits would become more and more eccentric until the matter moving in them would unite near one of the apsides with other portions of the ring. *We have thus a physical cause for the existence of this remarkable interval.*

CHAPTER XIII.

THE ASTEROID RING BETWEEN MARS AND JUPITER.

THE mean distances of the minor planets between Mars and Jupiter vary from 2·20 to 3·49. The breadth of the zone is therefore 20,000,000 miles greater than the distance of the earth from the sun; greater even than the entire interval between the orbits of Mercury and Mars. Moreover, the *perihelion* distance of some members of the group exceeds the *aphelion* distance of others by a quantity equal to the whole interval between the orbits of Mars and the earth. The *Olbersian* hypothesis of the origin of these bodies seems thus to have lost all claim to probability.* Professor Alexander's theory of the disruption of a primitive discoidal planet of great equatorial diameter, is less objectionable; still, however, it requires confirmation. But whatever may have been the original constitution of the ring,† its existence in its present form for an indefinite period

* A learned and highly interesting examination of this hypothesis will be found in a memoir "On the Secular Variations and Mutual Relations of the Orbits of the Asteroids," communicated to the Am. Acad. of Arts and Sciences, April 24th, 1860, by Simon Newcomb, Esq.

† For an explanation of the origin of the asteroids according to the nebular hypothesis, see an article by David Trowbridge, A.M., in Silliman's Journal for Nov. 1864, and Jan. 1865.

is unquestioned. Let us then consider some of the effects of its secular perturbation by the powerful mass of Jupiter.

Portions of the ring in which the periods of asteroids would be commensurable with that of Jupiter. — The breadth of this zone is such as to contain several portions in which the periods of asteroids would be commensurable with that of Jupiter. As in the case of the perturbation of Saturn's ring by the interior satellites, the tendency of Jupiter's influence would be to form gaps or chasms in the primitive ring.

The mean distance of an asteroid whose period is ½ that of Jupiter	. . . . .	=3·2776
That of one whose period is ⅓ of Jupiter's	.	=2·5012
" " ⅖ "	.	=2·8245
" " 2/7 "	.	=2·2569
" " 3/7 "	.	=2·9574
" " 4/9 "	.	=3·0299

For the purpose of facilitating the comparison of these numbers with the mean distances of the asteroids and of observing whether any order obtains in the distribution of these mean distances in space, we have arranged the minor planets, in the following table, in the consecutive order of their periods:

Periods and Distances of the Asteroids.

ORDER OF DISCOVERY.	NAME.	DISTANCE.	PERIOD.
8	Flora......	2 2014	1193 d
43	Ariadne............................	2 2034	1194·6
72	Feronia	2·2654	1245·4
40	Harmonia	2·2677	1247·3
18	Melpomene.........................	2·2956	1270·4
80	Sappho.............................	2·2971	1271·6
12	Victoria............................	2·3342	1302·6
27	Euterpe	2·3468	1313·2
4	Vesta	2·3613	1325·3
84	Clio..................................	2·3618	1325·8
30	Urania	2·3655	1328·9
51	Nemausa...........................	2·3657	1329·0
9	Metis	2·3858	1346 0
7	Iris	2·3863	1346·5
60	Echo	2·3931	1352·2
63	Ausonia	2·3949	1353·8
25	Phocea	2·4008	1358·8
20	Massilia	2·4144	1365·5
67	Asia..................................	2·4217	1376·5
44	Nysa.................................	2·4234	1378·0
6	Hebe.................................	2·4244	1379·0
83	Beatrice............................	2·4287	1382·5
42	Isis...................................	2·4400	1392·2
21	Lutetia	2·4411	1393·0
19	Fortuna.............................	2·4416	1393·5
79	Eurynome	2·4437	1395·3
11	Parthenope.........................	2·4519	1402·4
17	Thetis...............................	2·4737	1421·1
46	Hestia...............................	2·5262	1466·5
89		2·5498	1487·2
29	Amphitrite.........................	2·5544	1491·2
5	Astræa..............................	2·5772	1511·2
13	Egeria	2·5775	1511·4
14	Irene.................................	2·5860	1519·0
32	Pomona	2·5868	1519·6
91		2·5958	1527·5
56	Melete	2·5959	1527·7
70	Panopea............................	2·6129	1543·0
53	Calypso	2·6188	1548·0
78	Diana...............................	2·6236	1555·3
23	Thalia	2·6280	1568·0
37	Fides	2·6414	1570·0
15	Eunomia	2 6436	1572·6
85	Io	2·6466	1573·0
50	Virginia	2·6491	1575·0

Periods and Distances of the Asteroids—Continued.

ORDER OF DISCOVERY.	NAME.	DISTANCE.	PERIOD.
88	Thisbe	2·6553	1580·0
26	Proserpina	2·6561	1581·1
66	Maia	2·6635	1587·8
73	Clytie	2·6666	1590·5
3	Juno	2·6707	1594·2
75	Eurydice	2·6707	1594·2
77	Frigga	2·6719	1595·3
64	Angelina	2·6805	1603·0
34	Circe	2·6865	1608·3
58	Concordia	2 7014	1622·0
54	Alexandra	2·7123	1631·6
59	Elpis	2·7131	1632·3
45	Eugenia	2·7218	1640·1
38	Leda	2·7401	1656·8
36	Atalanta	2·7458	1662·0
71	Niobe	2·7501	1665·8
82	Alcmene	2·7547	1670·0
55	Pandora	2·7591	1674·0
41	Daphne	2·7657	1679·9
1	Ceres	2·7663	1681·0
2	Pallas	2·7696	1683·5
39	Lætitia	2·7740	1687·6
74	Galatea	2·7777	1690·9
28	Bellona	2·7785	1691·6
68	Leto	2·7836	1696·3
81	Terpsichore	2·8591	1765·7
33	Polyhymnia	2·8653	1770·6
47	Aglaia	2·8812	1786·4
22	Calliope	2·9092	1812·4
16	Psyche	2 9233	1826·0
69	Hesperia	2·9707	1871·1
61	Danaë	2·9837	1882·4
35	Leucothea	3·0040	1904·2
49	Pales	3·0825	1976·6
86	Semele	3·0909	1984·7
52	Europa	3·1000	1993·6
48	Doris	3·1094	2002·7
62	Erato	3·1297	2022·3
24	Themis	3·1431	2035·3
10	Hygeia	3·1512	2043·2
31	Euphrosyne	3·1513	2044 6
57	Mnemosyne	3·1565	2048·4
90	Antiope	3·1576	2049·4
76	Freia	3·3864	2276·2
65	Cybele	3·4205	2310·6
87	Sylvia	3·4927	2384·2

Remarks on the foregoing Table.

1. The first two members of the group, Flora and Ariadne, have very nearly the same mean distance. Immediately exterior to these, however, occurs a wide interval, including the distance at which seven periods of an asteroid would be equal to two of Jupiter.

2. On the *outer* limit of the ring Freia, Cybele, and Sylvia have also nearly equal distances, and are separated from the next interior member by a wide space including the distance at which two periods would be equal to one of Jupiter, and also that at which five would be equal to one of Saturn.

3. Besides these extreme members of the group, our table contains eighty-six minor planets, all of which are included between the distances 2·26 and 3·16; the mean interval between them being 0·0105. The distances are distributed as follows:

2·26 to 2·36	6	minimum.
2·36 to 2·46	19	maximum.
2·46 to 2·56	4	minimum.
2·56 to 2·66	16	maximum.
2·66 to 2·76	16	
2·76 to 2·86	8	
2·86 to 2·96	4	minimum.
2·96 to 3·06	3	
3·06 to 3·16	10	maximum.

The clustering tendency is here quite apparent.

4. The three widest intervals between these bodies are—

(*a*) between Leucothea and Pales........... 0·0785,
(*b*) " Leto and Terpsichore......... 0·0755,
(*c*) " Thetis and Hestia.............. 0·0525;

and these, it will be observed, are the three remaining distances, indicated on a previous page, at which the periods of the primitive meteoric asteroids would be commensurable with that of Jupiter. Now, if the original ring consisted of an indefinite number of separate particles moving with different velocities, according to their respective distances, those revolving at the distance 2·4935—in the interval between Thetis and Hestia—would make precisely three revolutions while Jupiter completes one. A planetary particle at this distance would therefore always come in conjunction with Jupiter in the same parts of its path: consequently its orbit would become more and more eccentric until the particle itself would unite with others, either exterior or interior, thus forming an asteroidal nucleus, while the primitive orbit of the particle would be left destitute of matter, like the interval in Saturn's ring.

5. If the distribution of matter in the zone was originally nearly continuous, as in the case of Saturn's rings, it would probably break up into a number of concentric annuli. On account, however, of the great perturbations to which they were subject, these narrow rings would frequently come in collision. After their rupture, and while the fragments were collecting in the form of asteroids, numerous intersections of orbits and new combinations of matter would occur, so as to leave, in the present orbits, but few traces of the rings from which the existing asteroids were derived. A comparison, however, of the elements of Clytie and Frigga shows a striking similarity; and Professor Lespiault has pointed out a corresponding likeness between the orbits of Fides

and Maia. For these four asteroids the nodal lines and also the inclinations are nearly the same; while the periods differ by only a few days. It is probable, therefore, that they are all fragments of the same narrow ring. Finally, as they all move nearly in the same plane, they must at some future time approach extremely near each other, and perhaps become united in one large asteroid.

CHAPTER XIV.

ORIGIN OF METEORS—THE NEBULAR HYPOTHESIS.

In regard to the physical history of those meteoric masses which, in such infinite numbers, traverse the interplanetary spaces, our knowledge is exceedingly limited. Such as have reached the earth's surface consist of various elements in a state of combination. It has been remarked, however, by a distinguished scientist* that "the character of the constituent particles of meteorites, and their general microscopical structure, differ so much from what is seen in terrestrial volcanic rocks, that it appears extremely improbable that they were ever portions of the moon, or of a planet, which differed from a large meteorite in having been the seat of a more or less modified volcanic action." As the celebrated nebular hypothesis seems to afford a very probable explanation of the origin of those bodies, whether in the form of rings or sporadic masses, its brief consideration may not be destitute of interest. We will merely premise that the existence of true nebulæ in the heavens —that is, of matter consisting of luminous gas—has been placed beyond doubt by the revelations of the spectroscope.

---

* H. C. Sorby, F.R.S.

As a group, our solar system is comparatively isolated in space; the distance of the nearest fixed star being at least seven thousand times that of Neptune, the most remote known planet. Besides the central or controlling orb, it contains, so far as known at present, ninety-nine primary planets, eighteen satellites, three planetary rings, and nearly eight hundred comets. In taking the most cursory view of this system we cannot fail to notice the following interesting facts in regard to the motions of its various members:

1. The sun rotates on his axis from west to east.

2. The primary planets all move nearly in the plane of the sun's equator.

3. The orbital motions of all the planets, primary and secondary, except the satellites of Uranus and Neptune, are in the same *direction* with the sun's rotation.

4. The direction of the rotary motions of all the planets, primary and secondary, in so far as has been observed, is identical with that of their orbital revolutions; viz., from west to east.

5. The rings of Saturn revolve about the planet in the same direction.

6. The planetary orbits are all nearly circular.

7. The *cometary* is distinguished from the *planetary* portion of the system by several striking characteristics: the orbits of comets are very eccentric and inclined to each other, and to the ecliptic at all possible angles. The motions of a large proportion of comets are *from east to west.* The physical constitution of the latter class of bodies is also very different from that of the former; the matter of which comets

are composed being so exceedingly attenuated, at least in some instances, that fixed stars have been distinctly visible through what appeared to be the densest portion of their substance.

None of these facts are accounted for by the law of gravitation. The sun's attraction can have no influence whatever in determining either the *direction* of a planet's motion, or the eccentricity of its orbit. In other words, this power would sustain a planetary body moving from east to west, as well as from west to east; in an orbit having any possible degree of inclination to the plane of the sun's equator, no less than in one coincident with it; or, in a very eccentric ellipse, as well as in one differing but little from a circle. The consideration of the coincidences which we have enumerated led Laplace to conclude that their explanation must be referred to the *mode* of our system's formation—a conclusion which he regarded as strongly confirmed by the contemporary researches of Sir William Herschel. Of the numerous nebulæ discovered and described by that eminent observer, a large proportion could not, even by his powerful telescope, be resolved into stars. In regard to many of these, it was not doubted that glasses of superior power would show them to be extremely remote sidereal clusters. On the other hand, a considerable number were examined which gave no indications of resolvability. These were supposed to consist of self-luminous, nebulous matter—the chaotic elements of future stars. The great number of these irresolvable nebulæ scattered over the heavens and apparently indicating the various stages of central condensation, very naturally suggested the idea that the solar system, and perhaps every other system in the universe, originally existed in a similar state. The sun was supposed by Laplace to have been an exceedingly diffused, rotating nebula, of spherical or spheroidal form, extending beyond the orbit of the most distant planet; the planets as yet having no separate existence. This immense sphere of vapor, in consequence of the radiation of heat and the continual action of gravity, became gradually more dense, which condensation was necessarily attended by an increased angular velocity of rotation. At length a point was thus reached where the centrifugal force of the equatorial parts was equal to the central attraction. The condensation of the interior meanwhile continuing, the equatorial zone was detached, but necessarily continued to revolve around the central mass with the same velocity that it had at the epoch of its separation. If perfectly uniform throughout its entire circumference, which would be highly improbable, it would continue its motion in an unbroken ring, like that of Saturn; if not, it would probably collect into several masses, having orbits nearly identical. "These masses should assume a spheroidal form, with a rotary motion in the direction of that of their revolution, because their inferior articles have a less real velocity than the superior; they have therefore constituted so many planets in a state of vapor. But if one of them was sufficiently powerful to unite successively by its attraction all the others about its center, the ring of vapors would be changed into one spheroidal mass, circulating about the sun, with a motion of rotation in the same direction with that of revolution."* Such, according to the theory of Laplace, is the history of the formation of the most remote planet of our system. That of every other, both primary and secondary, would be precisely similar.

In support of the nebular hypothesis, of which the foregoing is a brief general statement, we remark that *it furnishes a very simple explanation of the motions and arrangements of the planetary system.* In the first place, it is evident that the separation of a ring would take place at the equator of the revolving mass, where of course the centrifugal force would be greatest. These concentric rings—and consequently the resulting planets—would all revolve *in nearly the same plane.* It is evident also that the central body must have a revolution on its axis *in the same direction with the progressive motion of the planets.* Again: at the breaking up of a ring, the particles of nebulous matter more distant from the sun would have a greater absolute velocity than those nearer to it, which would produce the observed *unity of direction in the rotary and orbital revolutions.* The motions of the satellites are explained in like manner. The hypothesis, moreover, accounts satisfactorily for the fact that the orbits of the planets are all nearly circular. And finally, it presents an obvious explanation of the rings of Saturn. It would almost seem, indeed, as if these wonderful annuli had been left by the Architect of Nature, as an index to the creative process.

The argument derived from the motions of the various members of the solar system is not new,

* Harte's Trans. of Laplace's Syst. of the World, vol. ii., note vii.

having been forcibly stated by Laplace, Pontécoulant, Nichol, and other astronomers. Its full weight and importance, however, have not, we think, been duly appreciated. That a common physical cause has determined these motions, must be admitted by every philosophic mind. But apart from the nebular hypothesis, no such cause, adequate both in mode and measure, has ever been suggested;—indeed none, it seems to us, is conceivable. The phenomena which we have enumerated *demand* an explanation, and this demand is met by the nebular hypothesis. It will be found, therefore, when closely examined, that the evidence afforded by the celestial motions is sufficient to give the theory of Laplace a very high degree of probability.

A comparison of the facts known in regard to comets, falling-stars, and meteoric stones, seems to warrant the inference that they are bodies of the same nature, and perhaps of similar origin; differing from each other mainly in the accidents of magnitude and density. The hypothesis of Laplace very obviously accounts for the formation of planets and satellites, moving in the same direction, and in orbits nearly circular; but how, it may be asked, can the same theory explain the extremely eccentric, and in some cases retrograde, motions of comets and aerolites? This is the question to which we now direct our attention.

After the nuclei of the solar and sidereal systems had been established in the primitive nebula, and when, in consequence, immense gaseous spheroids had collected around such nuclei, we may suppose that about the points of equal attraction between the sun and neighboring systems, portions of nebulous matter would be left in equilibrio. Such outstanding nebulosities would gradually contract through the operation of gravity; and if, as would sometimes be the case, the solar attraction should preponderate, they would commence falling toward our system. Unless disturbed by the planets they would probably move round the sun in parabolas. Should they pass, however, near any of the large bodies of the system, their orbits might be changed into ellipses by planetary perturbation. Such was the view of Laplace in regard to the origin of comets.

It seems probable, however, that many of these bodies originated *within* the solar system, and belong properly to it. The outer rings thrown off by the planets may have been at too great distances from the primaries to form stable satellites. Such masses would be separated by perturbation from their respective primaries, and would revolve round the sun in independent orbits. Again: small portions of nebulous matter may have been abandoned as primary rings, at various intervals between the planetary orbits. At particular distances such rings would be liable to extraordinary perturbations, in consequence of which their orbits would ultimately assume an extremely elliptical form, like those of comets, and perhaps also those of meteors. It was shown in Chapter XIII. that several such regions occur in the asteroid zone between Mars and Jupiter. We may add, in confirmation of this view, that there are twelve known comets whose periods are included between those of Flora and Jupiter.

Their motions are all direct; their orbits are less eccentric than those of other comets; and the mean of their inclinations is about the same as that of the asteroids. These facts certainly appear to indicate some original connection between these bodies and the zone of minor planets.

The nebular hypothesis, it is thus seen, accounts satisfactorily for the origin of comets, aerolites, fire-balls, shooting-stars, and meteoric rings; regarding them all as bodies of the same nature, moving in cometary orbits about the sun. In this theory, the zodiacal light is an immense swarm of meteor-asteroids; so that the meteoric theory of solar heat, explained in a previous chapter, finds its place as a part of the same hypothesis.

CONCLUSION.

Some of the prominent results of observation and research in meteoric astronomy may be summed up as follows:

1. The shooting-stars of November, August, and other less noted epochs, are derived from elliptic rings of meteoric matter which intersect the earth's orbit.

2. Meteoric stones and the matter of shooting-stars coexist in the same rings; the former being merely collections or aggregations of the latter.

3. The most probable period of the November meteors is thirty-three years and three months. Leverrier's elements of this ring agree so closely with Oppolzer's elements of the comet of 1866 as to render it probable that the latter is merely *a large meteor* belonging to the same annulus.

4. The spectroscopic examination of this comet (of 1866) by William Huggins, F.R.S., indicated that the nucleus was self-luminous, that the coma was rendered visible by reflecting solar light, and that "the material of the comet was similar to the matter of which the gaseous nebulæ consist."

5. The time of revolution of the August meteors is believed to be about 105 years. M. Schiaparelli has found a striking similarity between the elements of this ring and those of the third comet of 1862. The same distinguished astronomer has shown,

moreover, that a nebulous mass of considerable extent, drawn into the solar system *ab extra*, would form a *ring* or *stream*.

6. The aerolitic epochs, established with more or less certainty, are the following:

1. February 15th–19th.
2. March 12th–15th.
3. April 10th–12th.
4. April 18th–26th.
5. May 8th–14th; or especially, 12th–13th.
6. May 19th.
7. July 13th–14th.
8. July 26th.
9. August 7th–11th.
10. October 13th–14th.
11. November 11th–14th.
12. November 27th–30th.
13. December 7th–13th.

About one-half of this number are also known as shooting-star epochs.

7. The epoch of November 27th–30th corresponds with that of the earth's crossing the orbit of Biela's two comets. The aerolites of this epoch may therefore have been moving in nearly the same path.

8. A greater number of aerolitic falls are observed—

1. By day than by night.
2. In the afternoon than in the forenoon.
3. When the earth is in aphelion than when in perihelion.

The first fact is accounted for by the difference in the number of observers; the second indicates that a majority of aerolites have direct motion; and the third is dependent on the relative lengths of the day and night in the aphelic and perihelic portions of the orbit.

9. The observed velocities of meteorites are incompatible with the theory of their lunar origin.

10. If the meteoric swarm of November 14th has a period of thirty-three years, Biela's comet passed very *near*, if not actually *through* it toward the close of 1845, about the time of the comet's separation. Was the division of the cometary mass produced by the encounter?

11. The rings of Saturn may be regarded as dense meteoric masses, and the principal or permanent division accounted for by the disturbing influence of the interior satellites.

12. The asteroidal space between Mars and Jupiter is probably a wide meteoric ring in which the largest aggregations are visible as minor planets. In the distribution of the mean distances of the known members of the group a clustering tendency is quite obvious.

13. The meteoric masses encountered by Encke's comet may account for the shortening of the period of the latter without the hypothesis of an ethereal medium.

APPENDIX.

A.

The Meteors of November 14th.

The *American Journal of Science and Arts* for May, 1867 (received by the author after the first chapters of this work had gone to press), contains an interesting article by Professor Newton "On certain recent contributions to Astro-Meteorology." Of the five possible periods of the November ring, first designated by Professor N., it is now granted that the longest, viz., 33¼ years, is most probably the true one. The results of Leverrier's researches in regard to the epoch at which this meteoric mass was introduced into the solar system, are given in the same article. This distinguished astronomer supposes the group of meteors to have been thrown into an elliptic orbit by the disturbing influence of Uranus. The meteoric stream, according to the most trustworthy elements of its orbit, passed extremely near that planet about the year 126 of our era; which date is therefore assigned by Leverrier as the probable time of its entrance into the planetary system. This result, however, requires confirmation.

Although the earliest display of the November meteors, so far as certainly known, was that of the year 902, several more ancient exhibitions may, with some probability, be referred to the same epoch. These are the phenomena of 532, 599, and 600, A.D., and 1768, B.C. (See Quetelet's

Catalogue.) The time of the year at which these showers occurred is not given. The *years*, however, correspond very well with the epochs of the maximum display of the November meteors. The intervals arranged in consecutive order, are as follows:

From B.C.	1768	to	A.D.	532,	69	periods of	33·319	years each.
" A.D.	532	to	"	599 5,	2	"	33·750	"
" "	599·5	to	"	902,	9	"	33·614	"
" "	902	to	"	934,	1	"	32·000	"
" "	934	to	"	1002,	2	"	34·000	"
" "	1002	to	"	1101,	3	"	33·000	"
" "	1101	to	"	1202,	3	"	33 667	"
" "	1202	to	"	1366,	5	"	32·800	"
" "	1366	to	"	1533,	5	"	33·400	"
" "	1533	to	"	1698,	5	"	33 000	"
" "	1698	to	"	1799,	3	"	33 667	"
" "	1799	to	"	1833,	1	"	34 000	"
" "	1833	to	"	1866,	1	"	33·000	"

The first three dates are alone doubtful. The whole number of intervals from B.C. 1768 to A.D. 1866 is 109, and the mean length is 33·33 years.

The perturbations of the ring by Jupiter, Saturn, and Uranus, are doubtless considerable. It is worthy of note that—

14	periods of	Jupiter	are nearly equal to	5	of the ring.
9	"	Saturn	" "	8	"
23	"	Uranus	" "	58	"

This group or stream has its perihelion at the orbit of the earth; its aphelion, at that of Uranus. (See diagram, p. 24.) It must therefore produce star-showers at the latter as well as at the former. Our planet, moreover, at each encounter appropriates a portion of the meteoric matter; while at the remote apsis of the stream Uranus in all probability does the same. The matter of the ring will thus by slow degrees be gathered up by the two planets.

B.

Comets and Meteors.

The recent researches and speculations of European astronomers in regard to the origin of comets and of meteoric streams, have suggested to the author the propriety of reproducing the following extracts from an article written by himself, in July, 1861, and published in the *Danville Quarterly Review* for December of that year:

"Different views are entertained by astronomers in regard to the *origin* of comets; some believing them to enter the solar system *ab extra;* others supposing them to have originated within its limits. The former is the hypothesis of Laplace, and is regarded with favor by many eminent astronomers. It seems to afford a plausible explanation of the paucity of large comets during certain long intervals of time. In one hundred and fifty years, from 1600 to 1750, sixteen comets were visible to the naked eye; of which eight appeared in the twenty-five years from 1664 to 1689. Again, during sixty years from 1750 to 1810, only five comets were visible to the naked eye, while in the next fifty years there were double that number. Now, according to Laplace's hypothesis, patches of nebulous matter have been left nearly in equilibrium in the interstellar spaces. As the sun, in his progressive motion, approaches such clusters, they must, by virtue of his attraction, move toward the center of our system; the nearer portions with greater velocity than the more remote. The nebulous fragments thus introduced into our system would constitute comets; those of the same cluster would enter the solar domain at periods not very distant from each other; the forms of their orbits depending upon their original relative positions with reference to the sun's course, and also on planetary perturbations. On the other hand, the passage of the system through a region of space destitute of this chaotic vapor would be followed by a corresponding paucity of comets.

"Before the invention of the telescope, the appearance of a comet was a comparatively rare occurrence. The whole number visible to the naked eye during the last three hundred and sixty years has been fifty-five; or a mean of fifteen per century. The recent rate of telescopic discovery, however, has been about four or five annually. As many of these are extremely faint, it seems probable that an indefinite number, too small for detection, may be constantly traversing the solar domain. If we adopt Laplace's hypothesis of the origin of comets, we may suppose an almost continuous fall of primitive nebular matter toward the center of the system—the *drops* of which, penetrating the earth's atmosphere, produce *sporadic* meteors; the larger aggregations forming comets. The disturbing influence of the planets may have transformed the original orbits of many of the former, as well as of the latter, into ellipses. It is an interesting fact that the motions of some luminous meteors—or *cometoids*, as perhaps they might be called—have been decidedly indicative of an origin beyond the limits of the planetary system.

"But how are the phenomena of *periodic* meteors to be accounted for, in accordance with this theory?

"The division of Biela's comet into two distinct parts suggests several interesting questions in cometary physics. The nature of the separating force remains to be discovered; 'but it is impossible to doubt that it arose from the divellent action of the sun, whatever may have been the mode of operation.'

"'A signal manifestation of the influence of the sun,' says a distinguished writer, 'is sometimes afforded by the breaking up of a comet into two or more separate parts, on the occasion of its approach to the perihelion. Seneca relates that Ephoras, an ancient Greek author, makes mention of a comet which before vanishing was seen to divide itself into two distinct bodies. The Roman philosopher appears to doubt the possibility of such a fact; but Keppler, with characteristic sagacity, has remarked that its actual occurrence was exceedingly probable. The latter astronomer further remarked that there were some grounds for supposing that two comets, which appeared in the same region of the heavens in the year 1618, were the fragments of a comet that had experienced a similar dissolution. Hevelius states that Cysatus perceived in the head of the great comet of 1618 unequivocal symptoms of a breaking up of the body into distinct fragments. The comet when first seen in the month of November, appeared like a round mass of concentrated light. On the 8th of December it seemed to be divided into several parts. On the 20th of the same month it resembled a multitude of small stars. Hevelius states that he himself witnessed a similar appearance in the head of the comet of 1661.'* Edward Biot, moreover, in his researches among the Chinese records, found an account of 'three dome-formed comets' that were visible simultaneously in 896, and pursued very nearly the same apparent path.

"Another instance of a similar phenomenon is recorded by Dion Cassius, who states that a comet which appeared eleven years before our era, separated itself into several small comets.

"These various examples are presented at one view, as follows:

"I. Ancient bipartition of a comet.—*Seneca, Quæst. Nat., lib. VII. cap. XVI.*

---

* Grant's Hist. of Phys. Astr., p. 302.

"II. Separation of a comet into a number of fragments, 11 B.C.—*Dion Cassius.*

"III. Three comets seen simultaneously pursuing the same orbit, A.D. 896—*Chinese records*—*Comptes Rendus*, tom. xx. 1845, p. 334.

"IV. Probable separation of a comet into parts, A.D. 1618.—*Hevelius, Cometographia*, p. 341.—*Keppler, De Cometis*, p. 50.

"V. Indications of separation, 1661.—*Hevelius Cometographia*, p. 417.

"VI. Bipartition of Biela's comet, 1845–6.

"In view of these facts it seems highly probable, if not absolutely certain, that the process of division has taken place in several instances besides that of Biela's comet. May not the force, whatever it is, that has produced *one* separation, again divide the parts? And may not this action continue until the fragments become invisible? According to the theory now generally received, the periodic phenomena of shooting-stars are produced by the intersections of the orbits of such nebulous bodies with the earth's annual path. Now there is reason to believe that these meteoric rings are very elliptical, and in this respect wholly dissimilar to the rings of primitive vapor which, according to the nebular hypothesis, were successively abandoned at the solar equator; in other words, that the matter of which they are composed moves in *cometary* rather than *planetary* orbits. May not our periodic meteors be the *debris* of ancient but now disintegrated comets, whose matter has become distributed around their orbits?"

C.

Biela's Comet and the Meteors of November 27th–30th.

At the close of Chapter IV. it was suggested that the meteors of November 27th–30th might possibly be derived from a ring of meteoric matter moving in the orbit of Biela's comet. Since that chapter was written similar conjectures have been started in the *Astronomische Nachrichten* * by Dr. Edmund Weiss and Prof. d'Arrest. The latter attempts to show that the December meteors may be derived from the same ring. The question will doubtless be decided at no distant day.

D.

The First Comet of 1861 and the Meteors of April 20th.

Recent investigations render it probable that the orbit of the first comet of 1861 is identical with that of the meteors of April 20th. The orbit is nearly perpendicular to the ecliptic.

* Nos. 1632 and 1633.

www.ingramcontent.com/pod-product-compliance
Lightning Source LLC
LaVergne TN
LVHW021415110826
845150LV00007B/1941

* 9 7 8 1 4 2 5 5 0 9 9 0 3 *